EFFECTIVE WRITING STRATEGIES

for Engineers

and Scientists

by Donald C. Woolston
Patricia A. Robinson
Gisela Kutzbach

LEWIS PUBLISHERS

Library of Congress Cataloging-in-Publication Data

Woolston, Donald C. (Donald Corlett), 1948-
Effective writing strategies for engineers and
scientists.

Bibliography: p.
Includes index.
1. Technical writing. I. Robinson, Patricia A.,
1948- . II. Kutzbach, Gisela. III. Title.
T11.W66 1988 808'.0666 88-589
ISBN 0-87371-099-1

LEWIS PUBLISHERS, INC.
121 South Main Street, Chelsea, Michigan 48118

PRINTED IN THE UNITED STATES OF AMERICA

Donald C. Woolston, a native of Montana, has been teaching technical writing in the University of Wisconsin–Madison College of Engineering since 1981. As an Adjunct Assistant Professor, he holds a joint appointment in the Department of General Engineering and the Pre-Engineering Office. His teaching efforts focus on a senior-level technical communication class for mechanical, industrial, and civil engineering students, but he has also helped teach freshman-level engineering courses and a special seminar on the Strategic Defense Initiative. In addition, he is the faculty advisor to the student organization that publishes the *Wisconsin Engineer* magazine. He has a special interest in computers, especially their use as a writing tool; he worked for IBM one summer writing manuals. In cooperation with Patricia Robinson and another colleague, he recently completed a 15-program videotape series on technical writing for engineers.

Professor Woolston's academic preparation includes a BA in psychology from the University of Montana, a PhD in physiological psychology from Duke University, and three years of postdoctoral training in the University of Wisconsin–Madison Department of Neurophysiology. His research into neural coding in the somatosensory and gustatory senses led to several publications in neuroscience journals, including the *Journal of Neurophysiology*. His interest in technical writing evolved from his interest in the communication aspects of neuroscience and his realization that he enjoys writing about research more than he enjoys the research itself.

Married with two teenage children, Professor Woolston lives in Madison, Wisconsin, where he enjoys running, tennis, basketball, and other sports that make the long Wisconsin winters more bearable.

Patricia A. Robinson is Adjunct Assistant Professor in the Department of General Engineering at the University of Wisconsin–Madison, where she teaches technical communication courses and directs the department's Technical Communication Certificate program. She also teaches seminars for technical writing practitioners and consults with industry. Professor Robinson is the author of *Fundamentals of Technical Writing*, and co-author of *Writing and Designing Operator Manuals*.

Professor Robinson did not expect to become a professor of technical communication. She graduated from Oberlin College in 1968 with a BA in Religion and from the University of Wisconsin–Madison in 1978 with a PhD in Hebrew and Semitic Studies. The market for Hebrew scholars was very soft in 1978, so she decided to draw on her experience as writer, editor, and writing teacher, gained from a variety of pay-the-rent jobs held during and between stints in graduate school.

In her spare time she is working toward an MS in Water Resources Management, which allows her to count the time she spends in a fishing boat as fieldwork.

Gisela Kutzbach teaches the major technical communication course designed for senior engineering students at the University of Wisconsin–Madison. For the College of Engineering's Technical Writing Certificate Program, she has developed courses in computer-assisted publishing and in graphical analysis and design using microcomputers. In addition she conducts in-house and on-campus seminars to help engineers and scientists improve their writing skills.

Professor Kutzbach was born in Berlin, Germany, where she received her BS from the Free University of Berlin. She holds an MS in meteorology and a PhD in the history of science from the University of Wisconsin. She has published extensively in the history of science, including a book on *The Thermal Theory of Cyclones* and, most recently, a historical analysis of the use of data graphics. During the past six years she has turned her attention from tracing the history of ideas and the thought patterns of scientists to tracing and improving the patterns and strategies engineers and scientists use to communicate their thoughts and findings. Professor Kutzbach is married and has three children.

Contents

Barriers in the Workplace
    The Writing Environment
    The Reading Environment
Targeting Your Writing
    Clarify Your Purpose
    Identify Your Audience
    Organize for the Reader
    Make Documents Readable
Team Writing
    Understand Group Dynamics
    Use a Three-Step Strategy
    Choose a Tactic
Summary

Basic Principles
    Moving from General to Specific
    Applying the Three Tell'em Theory
    Using Parallel Structure

Level Two: Combining Text and Graphics
Level Three: Publication-Quality Text and Graphics
Other Word Processing Products
The Computer Revolution in Writing: A Summary

Your Writing as a Part of a Legal Record
    What Is Discoverable Evidence?
    What Is the Role of the Engineer or Scientist?
Product Liability
    Legal Background
    Guidelines for Design-Related Writing
    Guidelines for Writing Instructions
    Guidelines for Writing Warnings
Construction Specifications
    Background
    Problem Areas in Specifications Writing
    Guidelines for Writing Specifications
Summary

Preface

I want to divide human understanding into two kinds—classical understanding and romantic understanding. A classical understanding sees the world primarily as underlying form itself. A romantic understanding sees it primarily in terms of immediate appearance.

Robert M. Pirsig
Zen and the Art of Motorcycle Maintenance

At least in terms of immediate appearance, the crafts of science and engineering are classical in that they require focusing on form, structure, analysis, and causation. They appear, at least at first glance, to be categorically different sorts of enterprises than writing, which might be called romantic since it requires intuition, imagination, and attention to feelings. Cold and calculating on the one hand. Intuitive and perhaps even whimsical on the other. Engineering, science, and writing seem as unlikely a mix as oil and water. Is it no wonder that some engineers and scientists have a hard time with writing?

Your authors, who make their living teaching scientists and engineers to write well, see a fascinating paradox in this unproductive, fatalistic approach to writing taken by some in science and engineering. We find that, paradoxically enough, the immediate appearance of what they write preoccupies many engineers and scientists—that is, they understand writing in a romantic sense. Successful writing requires, ironically enough, a classical approach, the approach that we engineers and scientists should be most comfortable with. We feel strongly that the only reason all scientists and engineers are not more analytic about their writing is that they do not have enough tools, enough knowledge, enough strategies, or enough insights to see beyond the superficial aspects of writing to its form and structure, which are the keys to excellence.

We wrote this short book on writing so that you, as an engineer or scientist, could have the insights and strategies necessary to be analytic about writing. After reading it, you should, with your new classical understanding, be able to write better in every respect: more quickly, more clearly, more convincingly. Our major premise is that anyone who is careful and analytical can write well when provided with some writing strategies and insights.

Perhaps you, like many, many of our students, are skeptical about writing ability and whether it can be learned at all. Perhaps you are thinking that writing skill is something that some of us were born with but you were not. And just perhaps you still believe that writing is so fundamentally different from other engineering and scientific activities that to be good at the first means not being good at the others. But the fact is that writing is design work, and engineers and scientists like you can and do design successfully. That is, you create something that fits a need by making sure that form fits the intended function. Writing well is nothing more than designing written documents that fit a need—and fit it well.

We admit that there are formidable obstacles to good writing that have to be overcome. The sobering reality about engineers' and scientists' writing is that it is time-consuming and exacting work. As one engineer we know put it, "Tech writing is fairly straightforward and is usually a simple application of common sense. My biggest problem is overcoming the chore of actually doing it. It is not fun." We see four particular problems:

1) Engineers and scientists are not well trained as writers. As one writing educator, George Douglas, admits:

> Very often, of course, writing is not served to scientists and engineers in a very palatable form. After all, in adolescence they have had to endure a number of years of exposure to the unimaginative ministrations of the schoolmarm grammarian, and their experiences in college . . . are often not a good deal more challenging. Freshman English . . . may be just another stale review of the principles of grammar, without much attention being paid to the

sorts of problems which actually plague most technical and scientific writing.

2) The writing that engineers and scientists do is difficult writing, since it requires conveying complicated information.

3) Unlike many aspects of these professions, writing is synthetic as well as analytic. It requires bringing ideas, concepts, and results together in addition to sorting them out, classifying them, or analyzing them. (Notice, however, that this is true of design work in general, not just writing.)

4) As a rule, engineers and scientists have the reputation of being inherently bad writers; some engineers and scientists seem curiously eager to see that reputation perpetuated.

This book addresses the first three of these problems. It provides further training in the sort of writing that engineers and scientists do; it gives specific advice for conveying complicated information; and it describes how to synthesize information according to specific writing strategies. (The fourth problem, the self-fulfilling proclamation about engineers and scientists being by nature bad writers, we dismiss out of hand. It simply runs against all our empirical evidence.)

Chapter 1 shows how on-the-job writing fits into the work of an organization. It addresses the conditions under which engineers and scientists must write, contrasting on-the-job writing with the writing learned in school. Most important, it gives you strategies for taming your writing tasks: analyzing your audience, identifying your purpose and stating it clearly, and dealing with team writing assignments.

Chapter 2 gives you insight into effective organizational strategies. By following the advice for treating writing similarly to other design tasks, you will be able to produce documents that do what they are intended to do—and you will write them in less time.

Chapter 3 encourages you to put thoughts into words quickly. It stresses that to write efficiently you must draft your documents as quickly as possible, but revise carefully and thoroughly. The chapter closes with specifics on troubleshooting your writing.

Chapter 4 shows you how efficient writers make liberal use of charts and graphs and suggests strategies for designing effective visuals.

Chapter 5 addresses the use of computers in writing. It includes information on desk-top publishing—using computer technology to produce documents that only a professional print shop could have produced a decade ago.

Chapter 6 discusses the legal aspects of writing. Most engineers and scientists must either write or approve documents that have legal importance, such as contracts, specifications, test results, and instruction manuals. It outlines some of the problem areas in writing and suggests strategies to help protect yourself and your organization from unnecessary litigation.

As a whole, the book reflects our goal of creating a useful writing guide for practicing scientists and engineers. It is focused directly on the needs of those in industry, government agencies, and consulting firms who must write about technical matters as part of their jobs. In our experience, this audience benefits most from hints, reminders, and a review of recently introduced writing tools, rather than an entire course in composition or rhetoric. This book represents our attempt at putting that information in one accessible volume.

Each chapter can be used independently from the rest of the book. Topics are liberally cross-indexed to allow quick retrieval. We rely on examples (both positive and negative) to illustrate concepts, but include no exercises, since the engineers and scientists we know don't lack writing practice—just writing help.

While the sections of the book each make separate points, the whole volume embraces two unifying credos of the authors. First, we believe that any scientist or engineer can write acceptable, useful technical reports and memos. Second, we believe that this ability depends more on analytic skills and a sincere interest in relating information to others than on being clever with words. The first credo arises from our extensive experience teaching eager but self-doubting engineering students to write better than they thought they could. The second arises from the nature of the writing. Technical writing is instrumental writing: a technical report is a tool for conveying information. Scientists and engineers are skilled designers; given the right training they can design with words, too.

Notice that we make no claim that writing can be made easy. Our hope is simply that this book makes writing as easy as possible for scientists and engineers.

REFERENCE

Douglas, George H. 1974. The common diseases of technical writing. *J. Technical Writing and Communication* 4:37–46.

Acknowledgments

Grateful thanks to Dr. John Beasley and Attorney Leslie F. Kramer for their comments on previous drafts of Chapters 5 and 6, respectively; of course, we the authors and not they are responsible for any errors that might remain in spite of their good help.

Chapter 1

Writing in the Workplace

Writing is the bridge between people and information. As part of your job, you generate information: sales projections, technical specifications, marketing policies, operating procedures. For this information to be useful, you must communicate it. Otherwise, you are like a book in an unknown language, full of interesting information no one can read. Sometimes the communication can be oral—a phone call or a meeting will be enough. But very often the communication needs to be in writing, for example, when a lot of people will be using it, when it will be used over a period of time, or when different persons need different parts of it.

If you have ever gotten the runaround on the telephone, you know how time-consuming it is to repeat a request for information or the details of a complaint. If many people need the information, a meeting may not be practical (conflicting schedules, different geographical locations), and even a meeting may require some backup documentation. If the information is to be used over a period of time, the only efficient approach is some sort of permanent record, usually in writing. And if different persons need different parts of the information, writing—on paper—is one of the only practical media, because it is accessible at any point—you don't have to fast-forward through the tape or scroll through the computer display to find the piece you need.

Because writing is a convenient means for communicating information, it is used extensively in organizations—to the dismay of some entry-level engineers and scientists. Once on the job, engineers and scientists spend about 30% of their time communicating, much of it in writing. As they advance in the company or agency, the communica-

1

tion percentage increases, until they spend more of their time communicating than doing engineering and science. To manage this function well, engineers and scientists must understand the context in which their writing is generated and used.

BARRIERS IN THE WORKPLACE

In an organization, what is written is influenced by the workplace environment. If ignored, the writing is likely to be unsuccessful: unclear, inappropriate, or, worst of all, simply not read. To be successful, the writer needs to overcome barriers to writing and reading presented by the workplace environment.

The Writing Environment

The environment for writing in the workplace is the polar opposite of what your high school teachers told you was necessary for writing a good paper. Study skills courses always counsel the student to block out plenty of time to do a project, avoid trying to do several different projects at once, and find a quiet place to study, secure from interruption. As high school students, we may have insisted that we needed the radio on to be able to think or that we worked best under pressure, but as working adults, most of us would give anything for a few hours of peace and quiet to write a report. The work environment seldom provides this. For most people, writing in the workplace means writing in spite of insufficient time, competing demands on the little time we do have, and constant interruptions.

On a typical day, you may receive several requests for information that have to be answered promptly, you may be working on two or three (or more) projects in different stages of completion, the phone may ring twenty times, half the people in the office may stop by to chat, and you may have to attend two or three meetings. When, in all of that, do you find time to write? If you wait for the uninterrupted block of time your study skills book prescribed, you'll never write a word. In addition, you have two problems that the high school student does not: you may not be able to get all the information you need before you start to write, and you may have to cope with sensitive issues of organizational politics.

The Reading Environment

The reader in the workplace endures many of the same obstacles as the writer: not enough time to read everything, competing demands, and constant interruptions. The difference is that the overburdened·writer will somehow produce something on paper, while the overburdened reader will *just not read everything.* This is the crucial fact, for it shapes the way effective workplace writers write. If your report is not read, communication does not take place, and you are back to being the book in an unknown language. If your report is read, it is because it beat out the competition—which is every other piece of paper on the reader's desk.

Writing winning reports is not as difficult or unlikely as it may seem, but because most people do not think about these environmental factors, most of those pieces of paper on the reader's desk are pretty uninviting. In the rest of this book we give you specific strategies to use to make your writing readable. The remainder of this chapter suggests ways you can target your writing to particular readers and discusses strategies and tactics for team writing projects.

TARGETING YOUR WRITING

Creating readable documents depends on the writer's willingness to take two steps before starting to write—or even to plan the writing. The writer must define the purpose of the document and identify the audience for it. Defining your purpose means more than just thinking to yourself: *It's a project report.* It means thinking through what that report will be used for—justifying recommendations, explaining expenditures, defending a design, or describing test findings. Identifying your audience means more than looking at whose name is on the letter of transmittal. It means figuring out who will actually use the report—usually many more persons than appear on the cover letter—and how the report will help them do their jobs. Here are some specific suggestions.

Clarify Your Purpose

Before you begin your outline, and certainly before you write a single sentence, think about why you are writing; that is, think about what the function of your report will be in your organization. If you are able, in specific words, to finish the sentence, "The purpose of this report is to . . ." you are ready to begin planning it. A good way to start is by writing a problem statement.

A good problem statement will help you as well as your reader because you can use it to help you decide what to include in your report. Once you know what the report is for, you can easily tell what information will help accomplish its aim.

A problem statement is one paragraph in which you tell the reader

- the reason for the project you're reporting
- the specific goal of the project
- the purpose of the report (you've already done this)

The first two parts set the scene for the reader. The sentences explaining the reason for the project give the reader a little background, a context for understanding the specific details of this particular report. They help the reader to fit this project into the activities of the organization as a whole. The sentences explaining the specific goal of the project help the reader to see the point of the information that you will present. The last part tells the reader what he or she is reading—and takes the burden off the reader to figure it out.

This three-part "formula" works for a problem statement for just about any kind of report. Here are several examples of problem statements from different kinds of reports.

Feasibility Study. Several years ago the ABC Corporation bought 65 acres of undeveloped land a few miles west of Central City. Since then the city has grown considerably and expanded to the borders of ABC's land. ABC hired our firm to determine whether it would be profitable to develop the land at this time into a residential section and shopping mall. We have completed our study and this report gives our recommendations.

Design Study. Masonville has outgrown its present wastewater treatment plant. The city is in violation of its discharge permit and faces stiff fines if the levels of suspended solids, biological oxygen demand (BOD), and heavy metals in the effluent are not reduced to permitted levels. My firm was asked to develop design specifications for a new treatment plant. This report explains our recommended design, and provides construction specifications and cost estimates.

Research Report. Wolves are active throughout the winter, often in subzero weather. While most of the wolf's body is well protected by fur, the pads of its feet are not. To find out why a wolf's feet do not freeze, even though it spends hours walking and running on snow and ice, we studied the effect of temperature on peripheral blood circulation. This report presents our findings.

All three of these opening paragraphs follow the three-part pattern.

The problem statement should be quite short and should appear right at the beginning of a document. The first few times you write one, you will probably be frustrated at the requirement that it be short. It will seem impossible to explain the reason for the project without explaining the background in some depth. Just remember that you can always go back and fill in the details in a section labeled "Background." All you want to do in the problem statement is to give your readers their bearings: *to tell them what your report is about and what it's for.* The remainder of your report will do the rest.

Identify Your Audience

Identifying your audience is composed of two sub-tasks: locating your readers and categorizing them.

Look Beyond Organizational Charts

Most organizations have tidy organizational charts showing exactly how all the job functions relate to each other and who reports to whom. Most organizations also do not actually operate so tidily. The actual lines of communication are much more complicated. In addition, most projects cut across organizational lines. A design for a new product, for example, may involve input from many different parts of your organization (or your client's organization): product development, marketing, engineering, production, and sales, as well as support services such as secretarial and typing help, art work, technical writing, accounting, and legal advice. Many people in many different areas will be affected by your work, and the readers for your report are likely to be extremely varied.

The best way to figure out who they all are is to think in terms of the project you are part of rather than where you or your client fit into an organizational chart. Think about who is involved with the project or who will be affected by it. Identify these people by their job functions. Once you have a sense of who they all are, you can begin to categorize them.

Decision Makers versus Advisors and Implementors

You can divide up your readers in many different ways, but one of the most useful is by how they will use your report. One set of readers will use the information in your report to make decisions; the other set will use it to advise on or implement those decisions.

For decision makers, the "bottom line" is the most important part. They want to know the conclusions and recommendations. Everything else is secondary, because their purpose in reading a report is to get information that will help them make a decision. They do not have time for a leisurely stroll through your reasoning or problem-solving process; they need to know the answer.

On the other hand, advisors and implementors need the backup information. They need to know what went into reaching your conclusion, and they need to know the details of implementing it. In a hierarchical organization, the ones at the top depend on subordinates to see that the information they receive is reliable and that their decisions are properly carried out. These people will read and digest the detail sections in your report, but usually only in their own areas of expertise. Thus, the engineer in charge of production will study the design specifications, the accountant will pore over the proposed budget, and the market analyst will consider the sales projections. But nobody will read your whole report, word for word, from beginning to end, except you and your typist.

Nonspecialists versus Specialists

Another way to divide your audience is into those people who are not technically sophisticated in your field and those who are. A civil engineer writing a design report for a wastewater treatment plant, for example, would write differently for a city council than for a city engineer. Interestingly, these two groups often match the first two: decision makers usually are not technical specialists, and the advisors/implementors usually are. This makes sense: a decision maker must be a generalist to be able to take into account all the different facets of a problem and to make a good choice, and a good advisor has to be enough of an expert to judge the quality of specialized work.

The major exception to these typical categories occurs when your audience is homogeneous, as, for example, when a biologist writes a paper for publication in a technical journal. Most of the time in organizations, however, the audience is mixed, and splits into these groups.

Organize for the Reader

Given that you have a mixed audience, how do you make a document useful to them? Succeeding depends on the twin tasks of selecting and arranging your information. You have to make sure that your readers have the right information to do their part of the job, and you have to make sure that they can find it. In a way, the rest of this book is about how to write for your audience, but the basic principles are these:

* Put yourself in your reader's shoes.
* Put the most important part first, then the next most important part, and so on.
* Put information in sections and label the sections.

Let us look at each of these in a little more detail.

Put Yourself in Your Reader's Shoes

Ask yourself how much the reader knows already and how much background you need to provide. Think about how detailed you should get in terms of what the reader would want. (Remember, unlike a college term paper, the object of workplace writing is to convey needed information, not to show how much you know.) Above all, always to remember that there is a reader, someone with whom you are trying to communicate, and someone who needs your help to do his or her job well.

Put the Most Important Part First

Most people, as we have already noted, are busy and short of time, so it makes sense to give the most important information "front page" space and save the less important parts for later in the report. That way, a busy reader can read only the beginning of a report, if that is all that time permits, and still not miss the major points. While this tactic makes sense, it is sometimes hard to put in practice. Too many of us remember our high school English teacher talking about leading up to a climax and keeping the reader in suspense. You are not writing a murder mystery; you need not wait until the last page to let slip the news that the butler did it.

Think of yourself instead as the newspaper reporter covering the murder—the headline would scream: BUTLER GUILTY! The analogy is fairly accurate. A newspaper has to convey its information efficiently in a minimum of space. Newspaper articles always put the essential facts at the top, and order the rest of the paragraphs by relative importance. That way, if space is tight, the editor can cut from the bottom and still be assured that the vital information will remain. Similarly, when you put the most important information in the beginning of a report, if time is short, a busy manager can cut from the end.

Put Information in Sections and Label the Sections

Because most of your readers will be interested in only part of your report, sectional organization makes it easy for them to find the parts they need. In other words, instead of trying to write a seamless essay, plan to write a report that is made up of discrete pieces linked together. Such a report is easier to write (especially in team writing situations, as we shall see later in this chapter) and it is easier to read. A reader can flip through the pages (or use the table of contents) to find relevant sections. Within a section, headings help the reader keep track of how the report is put together. Like the labels on file drawers and folders, headings and subheadings tell a reader what's "inside" and make it easy to move quickly and efficiently through a report.

Make Documents Readable

In the workplace, you cannot rely on a reader's being willing to struggle through prose a literary critic might call "difficult, but richly rewarding to the persistent reader." In the workplace, readers will not persist. No one has the time to read slowly, savoring the writer's turns of phrase (or cursing the writer's lack of clarity). The writer needs to make documents readable. Guidelines are

- Be concise.
- Highlight what's important.
- Focus on what the reader needs to know.

Let us look at each of these.

Be Concise

Chapter 3 addresses concision in more detail, providing specific guidelines, but a brief discussion is in order here. *Concision* means conveying your information in as few words as possible. It is not the same as brevity: a report can be a hundred pages long and still be concise—as long as there is a hundred pages' worth of information in it. Concision means finding the most efficient organization, the most economical phrase, and the most effective format for the particular audience and purpose you are trying to serve.

Highlight What's Important

For a reader with too much to read in too little time, help from the writer to see the essential points makes all the difference. Readers respond to cues like how much space a topic gets, how much detail it warrants, whether it appears at the beginning of a section or in the middle, and whether the writer has used visual signals, such as a heading or a list, to emphasize it. By highlighting what's important, you tell your reader what can be skimmed or omitted if time is short—and what can't.

Focus on What the Reader Needs to Know

Every writer has the twin tasks of selecting and arranging information. The key to selecting your information is to think about your subject from your reader's point of view, not your own. Think about how the reader will use the document you are writing, and think about what information is vital to that function—and then leave out everything else (or relegate it to an appendix). This may mean that you have to eliminate the portion that you most want to write about, but unless your favorite part will also be the reader's favorite part, so be it. Categories of information that often need to be cut are the history of a product, machine, or procedure; the theory behind an operation or design; and the step-by-step process by which you arrived at a conclusion.

TEAM WRITING

As if writing in the workplace were not already burdened with enough obstacles, you may frequently find yourself having to do some sort of team writing. Anyone who has ever served on a committee knows how difficult it can be to get a group of people to agree on a single decision, much less the wording of a report.

Yet most of us work as part of a team, working together on projects and somehow producing reports as a group. Much of the time the writing task seems less efficient and more time-consuming than the rest of the work put together, but it doesn't have to be that way. A few guidelines for group writing can help a great deal to make a team writing project relatively painless.

Understand Group Dynamics

Group writing, like any other group activity, is affected by the characteristics of the group. This is true whether the group is real or *de facto*. For example, a committee assigned to write a report of its activities is a real group. An individual writing a report whose work is reviewed and signed off by several other individuals is a *de facto* group—many of the same dynamics operate. In any group, these are some of the critical issues:

- lack of time
- control of the writing product
- ego investment in writing

How you deal with these issues will be the difference in whether your team writing projects are manageable tasks or major hassles. Let's look at them in more detail.

Lack of Time

A universal lack of time and a perception that groups are inefficient make most people reluctant and resentful about group writing projects. Meetings are seen as a waste of a scarce resource and an unwelcome interruption in the workday. The solution, of course, is to make the meetings worthwhile: short and efficient. To do this requires planning and leadership, but the rewards are well worth the extra investment of energy. In a *de facto* group, time becomes an issue when your reviewers are slow to sign off on your work, leading to delays that are not your fault, but for which you may take the flak. The solution to this dilemma is to make the review process simple and fast for the reviewer—which also takes planning.

Control of the Product

Power struggles over a variety of issues frequently develop in groups. In a team writing effort, such contests can be extremely disruptive and time-consuming when two sides battle endlessly over minor wording changes.

The key to minimizing the frequency and seriousness of these interruptions is twofold: giving group members autonomy and providing for frequent review in the development stages. These two may sound contradictory, but they really aren't. The first means that you give each group member a piece of the work that is his or hers alone—do not assign two group members to write one section together. The second means that having divided the work up, you also make sure that everybody is kept informed of what everybody else is doing all along the way.

Each person in the group, therefore, has both an investment of work in the final product and the opportunity to have a say about others' work. The first tends to keep a rein on the second: if I'm too critical of your work, you may retaliate by ripping mine. This method automatically helps to balance the power.

Ego Investment in Writing

You may have been told in grade school or high school that writing is self-expression. Well, it is; but so is everything else you do. Nevertheless, many people are a good deal more thin-skinned about their writing than they are about most other things. Such hypersensitivity can interfere with the inevitable adjustments and alterations that take place in meshing the work of several people into a unified, coherent report.

The best way to keep ego problems from holding up the work is to make as many as possible of those alterations and adjustments before the writing is in draft form. Most people are much more open to critique and suggestions when the writing is in outline form than when their ideas have been molded into sentences. We are less wedded to the organization of our ideas than to the expression of them in words—but the problem is that once the draft is written, you cannot change the order without changing the words. Paragraphs and sentences are linked together in a web of words that is not easily unwoven.

When you must edit a draft, or offer suggestions, do so gently, remembering that the writer probably feels very vulnerable. Word your criticisms to acknowledge the value in what the writer has done while at the same time suggesting an improvement.

Use a Three-Step Strategy

The key to successful team projects is planning, and this is especially true for a writing project. Good planning depends on leadership, but leadership is often slow to emerge in a writing project. Few people feel totally confident about their ability to organize their own writing effort, much less that of a group, so even natural leaders tend to sit silently waiting for someone to tell them what to do. The result is useless meetings and frustrated team members. This section outlines a three-step strategy for group writing that works. Once you learn it, you can guide the group and ensure that your meeting time will be productive and the report timely and well written.

The strategy we suggest combines group work and individual effort and is effective because it draws on the strengths of both. Groups are good at setting policy, but not at organizing information—that is a task best left to the individual. So our strategy has the group agree on common goals, but leave the actual writing tasks to individuals. It has three parts:

1. Group agrees on purpose, scope, and overall outline.
2. Individuals take assigned sections and develop detailed plans and write drafts.
3. Group reviews individual work frequently and provides feedback.

Group Decision Making

Just as an individual writer must define the purpose of a document, so must the group. The group as a whole, as well as each member individually, must share a common view of what function the report (or other document) will serve. The first time that the writing project is discussed, steer the conversation to the question of what the report is *for*—insisting that the group pin down the writing purpose at the start will save hours of wasted effort and talking at cross purposes later. Do not let anyone start discussing an outline, or even a division of tasks, until the group has a clear sense of purpose for the report.

To get the group to come to a decision may mean that you have to be persistent. A useful tactic that will keep you from being viewed as obstructive or obnoxious in your persistence is to feign confusion: "Wait, I still don't quite see the purpose. Could we just spell it out a little?" If people are, in fact, operating from different mind-sets, this tactic should make that obvious, so that the group can clarify its goals.

Once the purpose is clear, encourage the group to make decisions about the overall scope and shape of the report. As a group, develop a topic outline, blocking out the major sections of the report and agreeing on the main points to be covered in each section. In other words, guide the group to begin the basic writing tasks of selecting and arranging information. Do not go further than the basic outline, the table of contents, as it were, for the report. The more detailed work is better left to individuals, as suggested in the next step.

Individual Writing

As soon as the group has the outline, divide up the sections among the team members. If possible, get all members involved in some of the writing, even if it is only a short preface. That way everyone has a stake in the final product. You may well want to have one person serve as an overall editor, especially if the group is large, but even that person should do a little of the writing. The best way to divide things up is to let people volunteer for the parts they want to write, so that no one feels ordered around.

Exactly what happens next depends on the size and complexity of the report. If it's short and simple, people can just go to work writing drafts. If it is longer and more complicated, the individuals should develop detailed outlines and plans for their sections to bring back to the group before proceeding to write drafts. Comparing outlines helps avoid the problem of two people inadvertently covering the same point or nobody covering something essential.

Group Feedback

Try to keep the group members involved through the whole process. At regular intervals, have the members who are working on pieces of the report submit what they have to the rest of the group. You can tie the "status reports" to stages in the process, such as finishing a detailed outline, or simply make them weekly or biweekly. In either case, make sure they happen often, so that adjustments can be made before a lot of work is wasted. When members submit material to other group members (or when a solo writer submits material to reviewers), following these guidelines will help make the review process simple and efficient.

- Submit material early in the writing process.
- Request feedback often, but for small chunks of work.
- Request that comments be written.

Having members turn in material for group review early in the writing process—at the outline stage, for example—does two things. First, it ensures that problems in coverage or in approach to the subject can be caught and corrected early. Second, and perhaps more important, it helps to make people see writing as an iterative, trial-and-error design process, like any other kind of scientific or technical work, rather than as the result of a stroke of artistic genius. This adjustment in perception helps to keep the egos in check and makes people more receptive to critique.

Requesting frequent feedback keeps the members of the group in touch with each other and on task. In a *de facto* group, it keeps the reviewers involved and apprised of progress. If a reviewer has been exposed to a piece of writing all the way from the early outline stage to the final product, he or she is unlikely to demand major revision at the end—when changes are inconvenient and costly. By the time a reviewer looks at the final draft, he or she has already seen every piece in it several times. Keeping the review pieces small makes the review task manageable in a crowded workday, which in turn helps to keep the writing on schedule.

Finally, asking for written comments serves two purposes. The writer has a record of the changes that were requested—and the OKs that were given, making it easier to stand firm when asked for last-minute changes. It also limits the number of unnecessary changes. Some people feel that when they are asked to review a draft, they haven't done their job unless they change something, even if the change is slight. Having to put that change in writing tends to keep reviewers from cluttering the writer's task with trivial alterations.

The general strategy of having the group decide on the basic plan for the report, individuals do the actual writing, and all members of the group participate in reviewing and giving feedback is the most effective approach to team writing. Sometimes, however, certain members of the group are unwilling to participate actively or have real problems with writing. If that is the case, the best plan may be for the group to agree on purpose and scope, but for one or two individuals to do the bulk of the writing. This arrangement is a burden on the writer(s), but in the long run, it may be the most efficient. To work effectively, though, the group as a whole must take the time at the beginning to agree on the purpose and plan for the report. In addition to the general strategy outlined here, two specific writing tactics are especially well-suited to team writing, and these are outlined next.

Choose a Tactic

The strategy outlined above helps to get members of a group thinking about a writing project in the same way. The two tactics described in this section, modular organization and storyboarding, will help make the writing itself easier to fit together.

Modular Organization

Earlier in this chapter we suggested that you help the overworked reader by organizing your report in sections and then labeling the sections. Modular organization takes this approach a step further. The writing team chooses a *standard-length writing unit,* let us say two to five double-spaced manuscript pages, and then individual writers develop their outlines so that they can write drafts divided into units of that length. The units should be as self-sufficient as possible, so that a given module makes sense on its own, even if read out of context. You can be as flexible or rigid in the length requirements as you wish, but the modular approach has a number of advantages:

- The finished product appears uniform.
- The writers tend to write to the same level of detail.
- Modules are easy to rearrange.
- The writing and reviewing seem easier when the manuscripts are in "bite-sized" chunks.

All these help the team process.

If the finished product looks uniform, half the battle of blending manuscripts is won. One of the typical ways in which writers differ is the length of subunits in their writing. A report can end up looking very disjointed if one part has a major heading every three paragraphs and the next part has only one heading in as many pages. If the writers agree on a standard-length unit, such a disparity in format will not happen.

The difference is more than just in appearance: the amount of space devoted to a topic is a measure (in a competent writer) of the importance of a topic and therefore the level of detail it warrants. If you can get the group to agree on an outline and a standard-length writing unit, or module, you can tie the two together. In other words, you can agree that a certain level on the outline represents a module, and all writers working on their individual sections of the report operate on the same scale. The end result is a draft that reads in structure, at least, as if it were the product of a single hand. A good editor can usually smooth out the style differences to a large extent.

Part of team writing (or any writing, for that matter) is making changes. Even if the group agrees on an outline at the beginning, you may find along the way that you want to shift things around. If material is written in more or less self-sufficient modules, you can move them easily without leaving great holes or interrupting the flow of a larger unit. Even if you rearrange the presentation, you will still end up with a draft that gives a consistent impression to the reader. The other two advantages are mainly psychological. If you face the task of turning out many pages of report text, it is a lot easier to get started if you can think of it as a series of short, manageable modules rather than one huge, undivided mass. You can easily write three or four pages—but writing thirty seems quite daunting. Similarly, a reviewer asked to comment on a four-page draft can usually find time in the day to look it over and make some notes, when a thirty-page draft could get set aside for days while the reviewer tried to find a block of time to read it. As we have already seen, most people in the workplace do not find large blocks of time, only small pieces. Modular organization makes a small piece of time usable.

Storyboarding

The other useful tactic is storyboarding, which is a technique borrowed from scriptwriting. If you think of an outline as the specifications for a report and a draft as a prototype, the storyboards are the blueprints. They are a means of showing *more detail than an outline,* while retaining a form that permits easy change. A writer would complete a storyboard for every section of an outline covering more than a paragraph or two. In fact, if you are already using the modular structure, simply do one storyboard for every module. Figure 1 shows an example of a storyboard form. The basic elements in a storyboard are:

- statement of purpose
- list of main points
- summary statement
- sketch of any visual to be used

STORYBOARD PLANNING SHEET

WORKING TITLE *LAGOS PROPOSAL* AUTHOR *ROBINSON*

OUTLINE SECTION REFERENCE *II B*

PURPOSE OF SECTION *TO DESCRIBE SETTLING BASINS*

SUMMARY STATEMENT:

The plant will use 26 rectangular, horizontal-flow settling basins. Each basin will have a rapid-mix unit and flocculator attached and will provide mechanical sludge removal. Horizontal flow basins are better than vertical flow tanks because they have more capacity, are easier to operate, and are cheaper to build.

MAIN POINTS:

- USE RECTANGULAR, HORIZONTAL-FLOW BASINS: DESCRIP.
  ATTACHED RAPID MIX
  ATTACHED FLOCCULATOR
  MECHANICAL SLUDGE REMOVAL
- BETTER THAN VERTICAL-FLOW TANKS
  BIGGER CAPACITY
  EASIER OPERATION
  CHEAPER CONSTRUCTION

VISUAL:

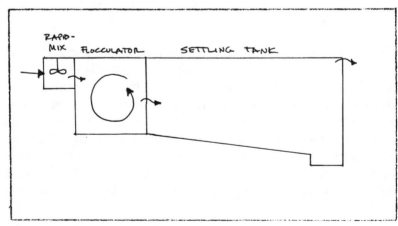

Figure 1. Example of a storyboard.

The arrangement of these may vary with the form used (storyboard forms are now available in fanfold computer paper), but each is important.

The statement of purpose simply tells what that particular section is *for*. Just as you need to define the purpose of the whole report before you begin to write (or even outline) it, you should do the same for each section. After all, you have selected and arranged your information consciously in a certain way: any unit, even down to the paragraph level, has a particular function to perform and a reason for being placed where it is. Stating the purpose overtly reminds you what those are and helps you remain focused as you write.

The list of main points provides a quick at-a-glance look at the scope of the section. They should correspond to what is already shown on the outline.

The summary statement tells in capsule form what the section says; it does *not* describe it. It does not say, for example, "This section discusses the fiscal implications of the proposed product mix." Rather, it might say, "The proposed product mix will boost daily profit by 40% while remaining within minimum production constraints." The summary statement is analogous to an informative abstract for a whole report. If you read the summary statement, you have the gist of the section—just not all the details.

The sketch of a planned visual can be anything from a diagram to a graph to a table—even a text table. A storyboard sketch gets you thinking about visuals and planning them from the beginning—which is the only way to end up with the marriage of verbal and visual communication you want. A good report makes the visuals seem integral to the presentation of information, not tacked on as an afterthought to liven up the presentation.

Like modular organization, storyboards help the team process by presenting a uniform appearance for easy review. In addition, they make it much easier to see overlapping coverage or gaping holes in the team's work than either outlines or drafts. Outlines do not give enough information, and drafts give too much. With an outline, you can see the structure, but you cannot see the form. With a draft, the reverse is true: the form and substance are there, but the structure is hidden beneath. Storyboards provide the quick check of how all the parts fit together while also giving a sense of how the report will read.

When the writing team meets to review progress, you can actually pin the storyboards up around the edge of the room and "walk through" the report. All the group members get a sense of how their work fits together, instead of the one appointed as editor being the only one with a sense of the whole. All the members can also offer feedback on where changes and adjustments need to be made. Instead of a power struggle between one editor and several writers, the process becomes one that elicits participation and cooperation from everyone.

SUMMARY

This chapter looks at how writing functions in the workplace as a bridge between information and the people who need to use that information. The workplace environment is not particularly conducive to either writing or reading, because of lack of time, competing demands, and other distractions, and because of that, the writer must take care to produce reports that can be used easily and efficiently. For busy readers, good reports are ones that are concise, that highlight important information, and that include only need-to-know information. For busy writers, the best way to write such reports is to begin by clarifying the purpose of the report and defining the audience for it.

Defining the audience requires locating and categorizing the audience for a report. Most readers fall into two major categories, those who make decisions and those who advise the decision makers and implement the decisions. Their information needs are different, but for both audiences, the best strategy for selecting and arranging the information in a report follows three rules: put yourself in the reader's shoes, put the most important information first, and put information in sections and label the sections.

Finally, this chapter explored the problem of team writing. Although similar in basic principles to solo writing, a writer in a team must work within the realities of group dynamics. These dictate an overall strategy for group work involving early agreement on purpose and direction, individual work on parts of the project, and frequent group review and comment. In addition, two tactics for group writing, modular organization and storyboarding, help to make the process work smoothly.

Chapter 2

Effective Organization

Any piece of writing will have a structure. However, if the structure of the work is *not* based on a plan, then it will reflect too much the personal preferences of the people involved. The difference between a planned work and an unplanned work is that *the planned work meets its purpose in an efficient manner.* For example, you can waste time finding a place in an unfamiliar city by asking passersby for directions or by following the general traffic pattern or by flipping a coin at each corner, or you can make that process more efficient by using a road map that has the route highlighted with a felt-tipped yellow marker. Similarly, you can waste time writing a memo or a report by following intuition or using a trial and error process. However, you can control the writing process by following a plan and using appropriate guidelines that reduce the amount of information processing, speed up the process of writing, and make reading more efficient.

Another advantage of planned structure is that it is repeatable. If a plan worked once, you can use it over and over again, or you can change it deliberately. Structure and design are just as beautiful and useful in writing as they are in technology.

This chapter covers organization guidelines that can have an immediate impact on your performance as a writer. It should give you confidence that efficient organization can be learned and that the extra work when organizing is a shortcut in disguise. The first section presents *three basic principles* for structuring information, which you can apply to any topic and to any scale of writing; the second section tells you *what to focus on when you develop your subject,* depending on whether you are describing, comparing, or persuading;

and the third section shows how *headings and other signposts* not only help you to structure information but also help your reader to grasp that structure and therefore to understand.

BASIC PRINCIPLES

Suppose you are planning to build a house. You tell your contractor your basic ideas, and then your contractor comes back with the blueprints. Now you can sit down and talk in earnest. In this situation, the blueprint parallels the functions and strategies of structure or organization in technical writing. We will briefly identify these.

You want to see the blueprint because it helps you to see if the contractor and you are thinking along the same lines, and if your ideas can be translated into reality. The blueprint provides a general picture of the new house. By just glancing at it, you know the exposure and the basic floor plan. It also allows you to think ahead. Seeing the arrangement of rooms, windows, stairs, and closets in outline form invites you to prearrange the furniture even before the house is built. And since the artist was consistent in using the same symbols for windows, stairs, and doors, you find it possible to "read" the blueprint, even though you are not an architect yourself.

The structure of your written documents serves functions and should incorporate strategies that are similar to those identified for the blueprint:

- You can make this structure visible as an external skeleton by using *headings and other signposts.*
- You can present *general statements before details* to orient the reader.
- You can *forecast and summarize* content and structure to prepare the ground for what is to come or to reinforce what has been said.
- Finally, you can aim at using appropriate and consistent *logical and grammatical structures* to help clarify meaning and relationships.

Just as one can use blueprints to portray anything from skyscrapers to a remodeled family den, so can one apply the organizing strategies discussed here to any scale of writing, be it an entire report, a segment of text, or only a paragraph.

Moving from General to Specific

Presenting general statements such as results or conclusions before the specifics such as observations, procedures, or descriptions of equipment has two distinct advantages: this strategy helps to clarify the significance of a message in the writer's mind and helps to answer the reader's question about why he should look at this piece of writing.

Bottom Line First

Have you ever tried reading a report without a title page *and* without introduction? Or can you imagine locating a book in an unfamiliar library that doesn't have a posted floor plan? Or can you recall a phone call at dinnertime when the person at the other end would simply not volunteer why he called? In any of these cases, you would be frustrated and feel as if someone were wasting your time. The reason is that you wanted the missing information right away so that you could use it, evaluate it, or disregard it; you wanted to be informed in order to stay in control.

On the other hand, suppose you are reading a detective story that starts with a minute description of the lonely hut in the moor and the traces of misdeed. A friend wants to volunteer "who did it," but you won't let him tell. The reason that you don't want to know the big picture right away is that you want to be entertained: you want to be kept in suspense and put the story together piece by piece like a puzzle. In a way, you want to relive the triumphs and trials of the detective himself.

These examples illustrate a basic difference between creative writing and technical writing:

Technical reports are not meant to entertain but are meant to inform, to persuade, or to motivate. The reader of such reports wants to know the "big picture" before the details.

The following situation illustrates this thought: Steve Beuer is an engineer working in a utility company. In an effort to save cost, he takes a close look at the cost of the streetlight program. He finds that the company's major supplier is switching from regular-style streetlights to power door–style streetlights. When the power door–style streetlight fails, just the power door assembly is changed; however, when the regular-style streetlight fails, the entire device is replaced. Suppose Steve has completed an economic study of the streetlights and sends this memo to his supervisor:

TO: Jay Friestadt, Engineer
FROM: Steve Beuer, Asst. Engineer
DATE: 13 April 1988
RE: Streetlight analysis

Our major supplier of streetlights intends to produce only power door–style streetlights after June 1, while our other two suppliers will continue with their production of regular-style streetlights. I have performed an economic analysis of the two alternatives. The following shows the results.

The troubleshooting crew reports that it takes 20 minutes less to repair a power door–style streetlight than to replace a regular-style streetlight. Regular-style streetlights require approximately 45 minutes for replacement.

In the 100 W and 250 W size, the regular streetlight has a 1–2% lower lifetime. In the 150 W size, the power door–style streetlight has a 1% lower lifetime.

In the past few years, the price of power door–style streetlights has dropped considerably. The 150 W size, which is most frequently used, costs now only $105.00, compared with $100.00 for a regular-style streetlight.

Power door–style and regular streetlights have approximately the same lifetime, and the initial cost for both of them is close to being equal. In addition, power door–style streetlights are cheaper to maintain than regular ones. Thus, the new streetlights are the more cost-effective choice.

I recommend that we switch to power door–style streetlights. Let's talk about this matter at our next Tuesday meeting.

Steve is quite happy with this memo because it shows that he has *considered the important factors for an economic analysis of the streetlights* and because *the memo highlights the facts that have led him to his recommendation.* He has written the memo the way he performed the work, in a problem-solving fashion—or as he would keep books: starting with the numbers, the facts, or the specifics, and ending with the bottom line, the conclusions, and generalizations. In fact, he proceeded in the chronological order of his investigation. He had heard about the new power door–style streetlights from personnel in maintenance and that led him to look into purchase costs and lifetime of the two alternatives. He is also proud that he has stated the purpose of the memo right in the first paragraph, thus orienting the reader.

At the other end, Jay Friestadt picks up the memo because he is waiting for some information on the streetlights. He's a bit annoyed with the subject line because it doesn't tell him that Steve has arrived at a conclusion on the matter. So he starts reading the memo to find out. Scanning the purpose statement, he knows that this is the memo he has been waiting for. The real question on his mind is what the utility company will have to do in this situation. Scanning through the details on labor, lifetime, installation, and maintenance costs, he tries hard not to overlook any important conclusion Steve might have come to. Since he does not like to be left waiting, he reflects: if only Steve would come to the point a little sooner. He's such a good worker, but why does he always have to tell me *what he did* before he tells me *what we should be doing?*

There you have it. Your readers, not only your supervisors but all your readers, want to know the bottom line first. As is shown in the revised version of the memo, *they'd rather read your conclusion and recommendation first.* Then, as they read through the supportive material, they can perceive the logic for including and arranging the rest of the information. And if they are pressed in time, they could quickly skim through the details or even skip them altogether.

Revision:

TO: Jay Friestadt, Engineer
FROM: Steve Beuer, Asst. Engineer
DATE: 13 April 1988
RE: Recommendation to change to power door–style streetlights

Our major supplier of streetlights intends to produce only power door–style streetlights after June 1, while our other two suppliers will continue with their production of regular-style streetlights. I have performed an economic analysis of the two alternatives. The following shows the results.

Power door–style and conventional streetlights have approximately the same initial cost and lifetime. In addition, power door–style streetlights are cheaper to maintain than conventional ones. Thus, the new streetlights are the more cost-effective choice.

I recommend that we switch to power door–style streetlights.

————————————————————— ✂

In the past few years, the price of power door–style streetlights has dropped considerably. The 150 W size, which is most frequently used, is now only $105.00, compared with $100.00 for a regular-style streetlight.

In the 100 W and 250 W size, the conventional streetlight has a 1–2% lower lifetime. In the 150 W size, the power door–style streetlight has a 1% lower lifetime.

The troubleshooting crew reports that it takes only 25 minutes to repair a power door–style streetlight but 45 minutes to replace a regular-style streetlight.

Let's talk about this matter at our next Tuesday meeting.

In this revised memo, the flow of information is from *general statements to specifics,* and the specifics are ordered in *descending order of importance.* Beginning a document, a section, or even a paragraph with general statements and then moving to specifics has distinct advantages. For your readers, it makes reading efficient; for you as writer it simplifies the writing process.

You can perform a simple test, the "scissors test" (Brunner, Mathes, and Stevenson 1980), to see if you have indeed structured your writing from generalizations to specifics. In the revised version of our memo, the scissors cut indicates that you could in fact cut the memo at that point and that the material above the cut would be sufficient for the reader to get the point or to make a tentative decision. This first component of the memo contains the *purpose, conclusions,* and *recommendations.* The format for structure here illustrated applies to all scales of writing. In a formal report, the first component constitutes the Executive Summary or management component.

Levels of Generalization

While specifics are factual and objective, generalizations are a matter of judgment. Specifics consist of empirical data and observations, such as the price of computers, or the repair times for power door–style and regular-style streetlights, or the number of pallets moved per hour. Such data derive from measurements, calculations, observations, or counts. Generalizations, however, occur at several levels: *results, conclusions,* and *recommendations* reflect increasing levels of generalizations and require an increasingly comprehensive ability to apply technical knowledge and experience in order to compare, interpret, evaluate, infer, or judge. The following example illustrates these levels of generalizations.

When I shop around for a home computer for the entire family and find the kind I want costs $995 in store A and $990 in store B, I would say that they cost about the same. I in fact interpret the data of my survey objectively by stating the **result** of simple comparing. I might then continue that both stores are within easy driving distance, but that maintenance service may be the real issue because the entire family is going to use the computer. Here I am making a deduction; that is, I am interpreting the results of my survey subjectively and state a **conclusion.** Finally, if I go on to say that I should buy the computer in store A because it is priced competitively and because the store has a good service department, then I am evaluating my conclusions in terms of the original issue of where to buy a computer. Now I am making a **recommendation.**

Thus, while results are considered objective statements requiring no subjective judgement, conclusions and recommendations reflect subjective interpretation of the results. Since results consist of averaging, comparing, or calculating, they are often grouped with sections on analysis, procedure, or discussion. However, because conclusions and recommendations form the bridge between technical investigation and the situation giving rise to the investigation—the organizational context—they are usually grouped prominently and separately from material on technical investigation.

A common mistake is not to distinguish clearly between these levels of generalization and to present results as if they were conclusions, or to present recommendations that are so open-ended that they do not relate well to the organizational context. Some examples follow:

Results presented as conclusions:	*Improvement:*
The rate of return for the proposed robotic system is 41%.	The proposed robotic system has a rate of return equalling 41%, which satisfies our company's 30% ROR requirement for investment implementation.

Unclear recommendations:	*Improvement:*
I recommend Design II.	I recommend Design II as the alteration of the current parking brake system design. Because this design consists of only six parts, it is cost-effective, easy to assemble, and easy to maintain.

Sometimes you may argue in defense that the facts speak for themselves, that the generalization or conclusion is obvious or implied. However, as a good writer you do not make your reader guess. Instead, you speed up your reader's information processing and show the maturity in knowledge and judgment that people expect from you by clearly identifying your subjective interpretations as conclusions and recommendations and by giving the reasons for these.

Ordering of Generalizations

A writer often deals not only with one generalization derived from specific data or observations, but with several. For example, a market analyst for a growing computer company is studying "where to establish additional distribution centers across the United States." His market analysis shows that profitable centers would be possible in seven cities. In his market analysis report he could list the seven cities in several ways:

- alphabetically
- by region
- by expected profitability
- by distance from headquarters

Or, an engineer proposes to change the drawing department of his company to the CADD system. In writing a proposal to this effect he would list the advantages for introducing the system. All of these advantages, of course, would be generalizations derived from the analysis of the needs of the graphic department, the capabilities of the CADD system, and other factors. The advantages could include time savings, cost savings, employee reduction, customer satisfaction, and employee happiness. When listing these advantages, he would naturally choose a descending order, which could reflect priority, difficulty, importance, or interest.

In either case, it would be important that the writer consciously choose an order that lets the reader know the logic of the ordering. Doing so will simplify the task of writing for the author and the task of understanding for the reader.

Writing Two-Component Reports

Keeping in mind that in technical writing we proceed from generalizations to specifics, we arrive at a basic outline that follows this sequence:

1. purpose statement
2. conclusions and recommendations
3. technical detail

This type of structure is like an *inverted pyramid*. Although the technical investigation and the specifics it generates form the basis for the conclusions, the pyramid is shown upside down to indicate that the conclusions and recommendations with their broad generalizations constitute paramount information for all readers of the document.

Figure 1 shows the common division of documents into *opening component* and *technical discussion*. You can follow this structure in documents of various length, in memos and letters as well as in reports. If you consistently use this format, you'll find that you reduce the amount of information processing both for your reader and for you as the writer.

In full-scale reports the opening component presents what is known as the Executive Summary, whereas the Technical Discussion brings the details of the technical investigation. The Executive

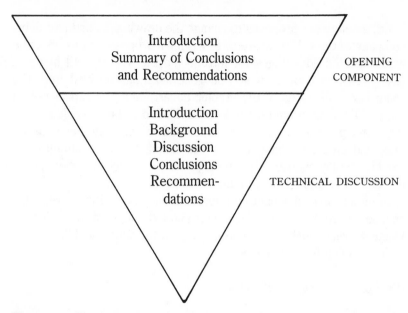

Figure 1. The inverted pyramid illustrates an effective method of structuring information for multiple audiences. While the opening component focuses on conclusions and recommendations, which interest all audiences, the technical discussion emphasizes the details of the technical investigation for experts and technical readers.

Summary interests all readers; however, because it gives generalizations and relates the work to the organizational situation, it is of special interest to the manager. In contrast, the Technical Discussion is geared more closely to the needs of the expert and technical audience. Because of the length of the full scale report, the Technical Discussion is written so that it can stand by itself, complete with an introduction, body, and ending. Figure 2 shows a sample outline for a long report.

The format clearly accommodates the needs of a mixed audience, as discussed in Chapter 1: the decision makers, often a nontechnical audience, and the advisors and implementors, often technical or expert people. We assume that the decision makers or managerial audience should not be inconvenienced and that they want up front the information most important for them, namely, the conclusions and recommendations and the reasons for these. Since this material also interests the experts and technical readers, they won't mind reading the Executive Summary (or the opening component) before going into the Technical Discussion, which will give them the information necessary to implement or follow the technical investigation.

REPORT OUTLINE

EXECUTIVE SUMMARY

OPENING
COMPONENT

Title or Subject Line
• Statement of the topic

For all readers, especially
managerial and nonexpert
readers

Purpose Statement
• Explicit statement of the organizational problem or situation addressed
• Questions or statement of task (indicating strategy and missing information)
• Purpose for writing

Conclusions and Recommendations

TECHNICAL DISCUSSION

Introduction to the Discussion

DISCUSSION COMPONENT

• Purpose statement (if long, formal report)
• Background

For expert or technical
readers

Technical Investigation
• Approach
• Scope of the work
• Criteria for evaluation

In short, informal reports,
such as memos, this
component provides only

• Methods
• Materials
• Techniques
• Design
• Data
• Findings/Results

material needed to support,
or fully understand, or
implement the claims in the
opening component.

Conclusions

In long, formal reports the
discussion component can
stand on its own.

Recommendations

APPENDIXES
Supplements
Exhibits

Figure 2.   General outline for a two-component report addressing multiple audiences.

As an engineer, you probably find it easier to write in the problem-solving format moving from specifics to general; however, you are also expected to write in the upside down way because *the decision makers want to know what to do before they want to know how you did it, what you saw, or even what it means.*

Writing Paragraphs

Moving from general to specific is not only an excellent guideline for writing anything from memos to reports, but it also is practical advice for writing paragraphs. Since paragraphs are the building blocks of writing, developing good paragraphs is desirable for the technical writer.

A paragraph has three functions: (1) it develops a unit of thought stated in the topic sentence; (2) it provides a logical break in the text; and (3) it creates a physical break on the page. The writer should make certain that the *topic sentence* is a general statement focusing on the key thought of the paragraph, while the bulk of the paragraph is specific information that supports the topic sentence. Moving on to the next general statement constitutes a logical break and requires development of that statement in a new paragraph. The white space between paragraphs makes this logical break visual and provides the reader with visual clues on how long he has to "flex his brains" for any one paragraph.

So, to a point, your reader should be able to get the gist of your message by merely reading the first sentence of each paragraph. Proofreading your drafts in this manner is a good check for you to see if your writing is coherent.

Applying the Three Tell'em Theory

Forecasting and summarizing content and structure is another technique to help the reader process information. The old preacher's saying proclaims that we should strive to be redundant in the following way:

Tell'em what you're gonna tell'em
Tell'em
Tell'em what you told'em

Repetition is one of the most powerful elements in teaching because it helps us to retain what we have learned. Similarly, repetition is a powerful tool of the technical writer. He uses repetition to

- present structure for the reader
- emphasize what is important
- reinforce understanding

The following explores each of these uses.

Use Repetition to Present Structure for the Reader

You can use repetition to forecast content and present structure at various levels of a document. You can use it to orient the reader, to help him decide what to read and what to leave out, and also to help him skim what you write. We can appreciate how this works when considering the function of the subject line in a simple memo. Consider this subject line from an internal memo in a utility company:

RE: Emergency equipment replacement and repair

The reader certainly knows right away the general subject of the memo but is left in the dark about its focus. In contrast, an effective writer makes the subject line forecast *the topic of the entire memo,* just as the topic sentence forecasts the topic of the paragraph. In this case the memo might have focused on one of a variety of topics:

RE: Evaluation of emergency equipment replacement and repair
RE: Procedures for emergency equipment replacement and repair
RE: Recent changes of emergency equipment replacement and repair
RE: Inadequacies of emergency equipment replacement and repair
RE: Financing of emergency equipment replacement and repair

As you will see, for the memo in question, the last subject line was correct, because it was the most helpful in getting the attention of the intended audience.

In the introduction of a document you have another opportunity to forecast the content and structure of the document. Chapter 1 showed specifically how to write an effective introduction by presenting first the reason, then the specific goal, and finally, the purpose for writing. This last point offers a chance to forecast both content and structure of the entire document. For example, the second introductory paragraph in Chapter 1 (see p. 5) ends this way:

> This report explains our recommended design, and provides construction specifications and cost estimates.

Thus, the reader knows the major points discussed in the report and their order of discussion.

The memo from the utility company had this introductory statement of purpose:

> RE: Financing of Emergency Equipment Replacement and Repair.
> Gas service employees identify unsafe, defective, or hazardous appliances and equipment in the customer's home. They have expressed concern that some customers may not make equipment repairs due to a lack of financing. We are looking for your help in determining if we need to develop a new financing program that could help to correct safety problems. **The following compares our present financing program with the program offered at Milwaukee Power and Light.**

After reading this introduction you expect that the body of the memo will be structured as a comparison of the two financing programs. By forecasting content and structure of the document you come across as organized, efficient, and caring.

When a document runs several pages, or when a report has sections and subsections, the writer still has other opportunities to forecast content and structure. At the beginning of each section, before going into the subsections, he can provide an overview of the section. So, instead of having two headings following each other without any text between them, the report section on the advantages of the CADD system begins as follows:

> 2.1   Advantages of a CADD system
> A CADD system has four distinct advantages for our operations: time savings, cost savings, employee reduction, and customer satisfaction. In the following we discuss these advantages in detail.
> 2.1.1   Time savings are crucial for us for several reasons. . . .

Similarly, brief overviews or lead-ins can effectively introduce lists. Not only will these lead-ins force the writer to consider if his lists are coherent (do all items indeed fit one topic), but they will also guide the reader.

And finally, the Three Tell'em Theory applies at the paragraph level. The topic sentence, which is a general statement focusing on the topic of the paragraph, should foretell the content as well as the structure of the paragraph.

The point is clear. By using the Three Tell'em Theory, you give structure in a series of "overlays," each of which presents overview before detail at the various levels or scales of the document and thus orients the reader.

Use Repetition to Emphasize What Is Important

We have a number of techniques to indicate what is important. The most common way to indicate importance is to discuss *first* what is most important. For example, if you wanted to persuade someone to use gas rather than electricity to heat a home, you would begin by discussing the cost advantage of gas (which is clear) rather than ease of installation or safety (which is debatable). We all, therefore, assume that unless we have other clues the first point in a list is the most important. Another way to indicate importance (that is, to emphasize a point), is to devote more *space* to discussing it than to discussing other points. If you devote three paragraphs to cost advantages of gas and only one to discussing availability and cleanliness of gas, you let the reader know your viewpoint. A third effective way to emphasize what is important is by *repeating* it.

By applying the Three Tell'em Theory, the writer touches on certain aspects of the message a minimum of three times. While forecasting structure draws attention to what is important, repeating and summarizing at the end of a paragraph, a section, or an entire document collects the reader's thoughts and thereby reinforces what has been said.

In addition to this general method, one can emphasize by *repeating selectively*. For example, an effective way to conclude a document is to restate the problem and objective before summarizing the solution. Also, since the conclusions are the most important aspect of a work, and since they document the writer's expertise as a problem solver, one emphasizes them by stating them both in the opening component of the document and at the end.

Another highly effective way to repeat selectively is to express the written text in *graphic form*. Showing a visual in addition to the text focuses attention on a particular point and singles it out from other points. Exhibits can be both graphical displays or word arrangements. For example, a table of numbers or a pie graph can summarize a cost estimate, while a table of words can emphasize the advantages of a CADD system. Chapter 4 gives guidelines on how to use visuals in technical communication.

Use Repetition to Reinforce Understanding

If you are concerned at this point that your writing might come across as boringly repetitious rather than as vigorous, consider a third important function of repetition. You, as a technical professional, present information that is not always easy to assimilate. In fact, you are often the one, perhaps the only one, who has worked in great depth on the problem or subject discussed in your writing. In order to *maintain speed of reading and understanding,* your readers will often welcome and need frequent reinforcement.

For the same reason, technical writing uses *repetition of terms rather than synonyms* of terms. Since the technical detail of installing a gas heater is complicated enough, the writer does not want to cause confusion by randomly interchanging terms like "procedure" and "process of installation."

Repeating material using different techniques of presentation also appeals to different learning styles. Most useful here is to *express difficult points both graphically and verbally.* Graphical displays, or visuals, can supplement and even echo the text. A reader might have had trouble following a verbal explanation of the greenhouse effect of the atmosphere but might have no trouble understanding the visual supporting the text.

Besides making the work of the reader easier, the Three Tell'em Theory will make writing easier: to present structure, you first have to establish structure; to emphasize what is important, you have to be clear about what is important; to reinforce understanding, you first have to understand. By forcing yourself to make these decisions in the planning stage and during the writing itself, you'll not lose track of your overall plan and objective.

Using Parallel Structure

So far we have discussed two strategies that are effective in structuring information: moving from general to specific and using the Three Tell'em Theory to forecast and summarize content and structure. These strategies work for all scales of writing, from entire documents to single paragraphs. *Parallel structure* is another strategy that is effective in producing logical documents and in clarifying meaning; it is also useful at the sentence level (see Chapter 3 and 6). Let's explore what this strategy is and how it works.

The thinking of engineers and scientists as well as writers is profoundly influenced by the concepts of *unity* and *symmetry*. The search for a unifying principle in nature led to discoveries such as the law of conservation of energy and the concept of species. Equally important were discoveries of symmetry: products of cell division are identical cells, and chromosomes come in pairs. The double helix of DNA strands, which parallel each other from one end to the other, is perhaps the ultimate example of symmetry. These concepts are also prominent in writing. Unity is integral to the structure of the paragraph. And symmetry in writing reveals itself in the powerful hexameters of Homer's *Iliad* as well as in the table of contents of a book.

Parallel structure is the most useful form of symmetry in technical writing. It mimics symmetry of *meaning* by symmetry of *structure* in order to clarify the information in the writer's and reader's mind. Parallel structure has these important functions:

- to clarify the logic of development in outlines
- to set up items for comparison and contrast
- to show that points are related
- to create a symmetry that appeals to the reader and enhances understanding

Outlines (and tables of contents) should be checked for internal consistency and flaws in parallel structure. You can check for parallel structure by drawing your outline like a family tree.

Using parallel structure is crucial for clarity and understanding when you set up things for comparison. For example, if you discuss three alternatives for reducing the cost of emergency hand brakes, the outline should reflect that you considered all three alternatives and that you applied the same criteria in each case. For the sake of symmetry, you cannot omit discussion of one criterion in one case simply because it does not apply. You can discuss it by saying that it does not apply. This kind of symmetry becomes particularly apparent when you use graphics such as a table.

Lists by definition combine points or items that are related. Within a paragraph, lists must rely on grammatical symmetry, while lists set off from the rest of the text use visual symmetry in addition. In either case, parallel structure immensely aids in understanding. In the following example in the left-hand column, each sentence is grammatically correct, but reading is unnecessarily cumbersome because of shifts in subject from you to other points of view, shift from imperative to the active and passive voice (see Chapter 3, p. 58), and unclear identification of list items. The revised version in the right-hand column is much clearer and aesthetically appealing because it applies parallel structure. In addition, it applies the Three Tell'em Theory in that the topic sentence prepares the reader that the procedure has four steps. (Of course, other versions are possible; the point is that you must be consistent. As Chapter 6 shows, consistency will pay off especially in formatted lists such as procedures.)

Original:	*Revision:*
The process of home weatherization follows a well-defined procedure. First, begin preparations by ranking weatherization groups, with 3–4 unit dwellings having top priority. You must also identify potential candidates based on landlord information and tenant payment history. Sending letters explaining the weatherization project to qualifying landlords and making follow-up calls is another step. Finally, a home energy checkup must be performed and possible weatherization measures must be identified.	The procedure of home weatherization follows a well-defined 4-step procedure: (1) Begin preparations by ranking weatherization groups, with 3–4 unit dwellings having top priority. (2) Identify potential candidates based on landlord information and tenant payment history. (3) Send letters explaining the weatherization project to qualifying landlords and make follow-up calls. (4) Perform a home energy checkup and identify possible weatherization measures.

In general, we are much more sensitive to visual symmetry than to logical symmetry, simply because in this case the asymmetries stick out at a glance. It is always especially worthwhile to check for consistency in the labeling and typography of headings.

In this section we have discussed three powerful strategies for structuring information: moving from general to specific, using internal forecasts and summaries, and using parallel structure. These strategies are useful during the writing process to make the information more accessible to the reader. We will now turn to specific techniques that not only help *reach an audience* but also help *clarify the purpose* of a message.

PICKING A STRATEGY FOR DEVELOPING SUBJECTS

Although deciding how to develop a certain subject, such as the description of a tractor, the comparison of two software packages for graphics, or the instructions for setting up a tent, is affected by our own thinking on the subject, its development should be a carefully considered design choice. For example, when considering the construction of a highway, the department of transportation has a choice among blacktop, asphalt, and cement surfaces. The decision which surface is the best depends on the function of the highway and local circumstances. Similarly, one can simplify the task of developing a subject in writing by choosing a standard pattern of development that suits the purpose and audience of the message. The patterns we discuss fall into three groups:

- focusing on particulars
- focusing on time sequences
- focusing on conclusions

We will explore how and when to apply these basic patterns.

Focusing on Particulars

In certain writing tasks, even though you are considering a complete subject, such as the description of a manufacturing process or a lawn mower, you develop the subject by analyzing or describing its parts. In such cases, you are developing a whole by focusing on its specifics.

Whole/Parts Analysis

Suppose that you have to write an article on power plants. You could proceed either by analyzing the branches or divisions of, for example, a coal power plant, or by classifying coal as one of many sources of energy for public use.

If you analyze the power plant's branches and divisions, you are focusing on its parts; for example, you would go on to explain the distribution system, the generation of electricity, or the system for pollution control. If you classify coal as one of many sources of energy, you would view the coal power plant as part of a larger category and group it with nuclear power plants, hydroelectric plants, and natural gas.

In the first case, you move down the ladder of abstraction, so to speak, and in the second you move up, but in each case you focus on parts or specifics rather than a whole. Whole/parts analysis is a good starting point for the development of many subjects. It not only helps explore a subject, but also can provide the basic structure for a paragraph or an entire report.

SAMPLE OUTLINE *for whole/parts analysis:*

- State the whole
- Explain the parts
     use decreasing order of importance or
     other logical subpattern

Technical Description

Technical descriptions, which are tailored to specific audiences and are often part of promotional material, manuals, brochures, and reports, incorporate whole/parts analysis. The following outline, adapted from the whole/parts outline, provides some general guidelines for writing effective technical description.

SAMPLE OUTLINE *for technical description:*

- Definition
     informal or formal definition of object
     function of object

- Physical description
  general appearance
  break down into parts or units and their functions
- Operational description
  how all parts work together or
  one operational cycle

These three parts of the technical description answer three questions:

What is it?
What does it look like?
How does it work?

The following shows how to give the answers.

When defining an object or a process, you can use four different methods. An informal but highly effective method, especially for a general audience, is to use an *analogy,* that is, to define something that is unfamiliar by something that is familiar. For example, famous analogies in the history of science are James Jeans's comparison of light waves with water waves or Clark Maxwell's analogy between lines of force and streamlines in an incompressible fluid. In other cases a simple *synonym* will do, such as that sodium chloride is salt. Further, a widely used approach is to define an object or a process by *summarizing its function,* as in this example: "A mousetrap is a spring-loaded trapping device used to catch small rodents." In our fast-paced society, readers often prefer this type of definition because they are highly interested in things like purpose, function, or use. Finally, if you want to emphasize how something differs from similar things, you may use a *formal definition,* in which you identify the broad category to which something belongs and then show the distinguishing characteristics. An example is: "Nitrogen is a gas with the atomic number seven."

The physical description of an object emphasizes its parts or components and their relationship to the whole. Applying the overall strategy of moving from general to specific, you begin by describing the general appearance and characteristics, such as material, dimensions, weight, price, lifetime, and use—and you can supplement this section with a visual of the whole. Next, following the same pattern of providing an overview before going into detail, you focus on the description of the parts. You can partition the parts by location within the object, importance, assembly, or function; and you can supplement the discussion of the parts by using visuals, such as cutaway views, exploded views, or views of the parts alone.

The final section of the description explains how the parts fit together and work together. In short descriptions, this section is also the concluding paragraph.

Focusing on Time Sequences

A whole group of writing tasks requires the writer to structure information as a sequence of steps, often defined by time. These are process explanations and directions and procedures for actually conducting a process. Moving from general to specific, we arrive at the following general outline for these types of technical communications:

SAMPLE OUTLINE for a process or procedure explanation and directions:

- State the purpose of the process or procedure and identify basic stages or phases
- Describe the process or procedure step-by-step using chronological order

Process explanations are not only part of technical descriptions, which were discussed in the previous section, but they also occur independently in sales literature, manuals, reports, or textbooks and are intended for many different audiences, including managers, technicians, and the general public. Process explanations are not meant to enable a reader to actually perform a certain task; rather, they are meant to provide a sufficiently complete understanding of the process. For example, a process explanation of "adding oil to the crankcase of a lawn mower" would be:

Before each use of the lawn mower, the operator must check if the oil level is SAFE. If the oil level is low, only enough oil should be added to raise the level to the FULL mark on the dipstick.

In contrast, the corresponding set of directions would be:

Overview and Purpose:	Before each use, assure oil level is in SAFE range. Add oil if level is low.
Step-by-Step:	1. Position mower on level surface and clean around oil dipstick.
	2. Remove dipstick by rotating cap counterclockwise 1/4 turn.
	3. Wipe dipstick and insert it into filler neck. Rotate cap clockwise 1/4 turn. Then remove dipstick and check level of oil.
	4. If level of oil is low, add enough oil to raise level to the FULL mark on the dipstick. **Do not fill above the full mark, because the engine could be damaged when started.**
	5. Insert dipstick into filler neck and rotate cap clockwise 1/4 turn to lock.

It is clear that the set of directions to actually perform the task contains considerably more detail than the description of the process.

Process explanations are written in the *indicative mood*; the active voice is preferred, although passive voice is not always avoidable. The operator can be referred to but is never directly addressed. Thus, process explanations are commonly written in the *third person* (he, she, one, it, they). The emphasis is on how it works, not on how to do it.

In contrast, sets of directions directly address the operator or reader; thus they are usually written in the *imperative mood* (wipe the dipstick). The process is broken down into small steps, each describing a single required action. Emphasis is on the precise details of how to do it. Chapter 6 discusses in detail common flaws in writing instructions and suggestions for avoiding them.

Because process explanations intend to provide an overview and often incorporate some theory to enhance understanding, they often profit from simplified *visual presentations*. For example, a simple *flowchart* of the painting process of window-blind hardware reinforces understanding, even though important technical information is omitted. Similarly, a simple *time line* illustrating the schedule for converting inventory procedures to computers is easily understood and appreciated by a wide range of readers.

Focusing on Conclusions

The purpose of the two basic patterns of development discussed so far has been to highlight specifics and to provide details on parts, stages, or steps. Although the organization of these patterns moves from whole to parts, general to specific, or overview to detail, the emphasis is on the detail rather than the whole. In the third kind of pattern of development the emphasis is exactly opposite. Patterns discussed in this section focus on conclusions; details are no longer ends in themselves but serve only as support for these conclusions. We can distinguish three different patterns that have this emphasis:

persuasion
comparison/contrast
cause/effect or effect/cause

All these patterns follow a basic and very obvious outline:

- State the claim
- State support for the claim
    use descending order of importance

Persuasion

You could argue that all writing is basically meant to be persuasive. Even if you think that you are only informing, as in the technical description of a mouse trap, you still want to persuade a reader that the information you provide is correct. However, in such cases persuasion is not the dominant pattern of development.

In contrast, persuasion is the dominant purpose of communication when you present a proposal, or solutions to problems, or recommendations. In these cases, your writing will be most effective if you use persuasion as the appropriate pattern of development.

To be convincing, you should present your *strongest arguments first.* In most cases, you are going to be more effective and efficient if you state the claim up front and then list your support than if you wait until the end of the message to come up with the punch line. Similarly, you show confidence in your solution by stating the support in *descending order of importance* rather than reserving the strongest argument for the end. In fact, in presenting support for your claims

ım *positive to negative* (that is, from advantages to
ɔu can anticipate and rebut):

for persuasion:

ɪrt
ˑ of importance
bjections

• Restate claim

This approach is straightforward, shows your viewpoint, and is, above all, convincing. For example, assuming that you have to show why your office should adopt a certain word processing system, you could begin by discussing its superior speed, comprehensive built-in spelling checker, and other desirable features before explaining that the lack of footnoting capability is not important for your office.

Only in the most sensitive cases and issues, when you know that your reader does not agree with your claim (that he or she has already a formed opinion on the subject), do you proceed in a more roundabout way. Let's assume that the workers in the assembly line that you want to partially automate are afraid of losing their jobs. In that case, starting with praises of the advantages of automation is not going to persuade them. Rather, you do better if you show that you understand their needs and respect their opinions. You begin by describing the details of the present assembly line, outlining its advantages and then disadvantages. Then you can gradually work into explaining the automated assembly line, guiding and leading your audience in a discussion that they now might be willing to follow. Again, deciding on the approach depends on your audience analysis. If no problems exist, the sample outline above is your best choice.

Comparison/Contrast

As with all patterns of development, comparison/contrast can provide the basic structure for an entire report or for only a paragraph. It is especially useful when accompanied by a conclusion, the interpretation of the comparison itself. Because a complete comparison requires an interpretation, it involves a strong persuasive element. The basic outline therefore closely follows the outline above for persuasion:

SAMPLE OUTLINE for comparison/contrast

- State conclusion about the relationship of things being compared
- Provide overview (of points being compared)
- Provide point-by-point analysis of differences and similarities
    descending order of importance
    positive to negative

Pattern A			Pattern B		
1. Subject (1)			1. Point (1)		
	point	(1)		subject	(1)
	point	(2)		subject	(2)
	point	(3)	2. Point (2)		
2. Subject (2)				subject	(1)
	point	(1)		subject	(2)
	point	(2)	3. Point (3)		
	point	(3)		subject	(1)
				subject	(2)

- Restate conclusion

The outline shows that the crucial information in this pattern is not merely the facts, which constitute the comparison/contrast itself, but the conclusion derived from the point-by-point analysis. A comparison/contrast without interpretation wastes time because it forces the reader to do the work of the author. To illustrate, let us look at the following example from *The Practice of Management* (1986, 194) by Peter Drucker, whose immense success we can attribute in part to his clear writing:

Most managers, especially in larger companies, have learned the hard way that performance depends upon proper organization. But the practical manager did not as a rule understand the organization theorist, and vice versa. . . .

We know today that when the practical manager says "organization," he does not mean the same thing the organization theorist means when he says "organization." The manager wants to know what kind of a structure he needs. The organization theorist, however, talks about how the structure should be built. The manager, so to speak, wants to find out whether he should build a highway and from where to where. The organization theorist discusses the relative advantages and limitations of cantilever and suspension bridges. Both subjects can properly be called "road building." Indeed, both have to be studied to build a road. But only confusion can result if the question what kind of a road should be built is answered with a discussion of the structural stresses and strains in various types of bridge.

This brief excerpt starts with a claim, goes on with an *overview* of the comparison, uses an *analogy* to clarify meaning, gives examples as support, and ends with a *restatement* of the claim. Throughout the passage, Drucker has used parallel structure to enhance clarity.

Since *feasibility studies* often involve comparing various alternatives, the good writer does well to structure the discussion of his study accordingly. In this case, the point-by-point comparison revolves around predetermined criteria. A visual such as a word table can provide a useful overview of such a comparison. The general outline follows:

SAMPLE OUTLINE for a feasibility study:

- Present purpose statement
- Present background
- Provide overview of recommendations
- State criteria to evaluate possible solutions
  select criteria that will identify significant differences between solutions
- List possible solutions
  select only solutions that are technically feasible
- Provide detailed comparison/contrast of solutions
  on the basis of the criteria selected
  Use either Pattern A or Pattern B. In Pattern A, order from most to least desirable solution (anticipate your conclusions). In Pattern B, order from most to least desirable criterion.
- State conclusions
- State recommendations in detail

Cause/Effect

This pattern again focuses on the conclusion, a claim that a certain cause/effect relationship exists. After stating the *question* that defines the causal relationship to be determined, the writer goes on to state the *conclusion*. The next step is to support the conclusion with specific *evidence*. The support either traces the sequence from cause to effect or lists the causes or effects in descending order of importance.

SAMPLE OUTLINE for cause/effect:

- State the question (if necessary)
- State the claim
    state the cause/effect relationship
- Provide the support
    explain logical cause/effect sequence or
    begin with most important cause or
    begin with most probable effect
- Anticipate and meet objections

Applying the basic patterns of development presented here and modifying them for specific situations should help to reduce the time and effort of drafting and revising. We have presented these strategies in three basic groups, each with a different focus, and we have explored their use individually. However, the writer uses these strategies more often in combination than separately. For example, a comparison of two word processing systems can include a process description of "using the spell checker" for each system and a whole/parts analysis of each according to screen menus.

Whatever strategy you use in structuring your writing, your readers will benefit if you let them know what you are up to. A convenient way to forecast content and structure for your readers is to use appropriate headings and other signposts.

HEADINGS AND SIGNPOSTS

Since technical documents are working documents, most readers will only read parts of them or refer to parts of them. The majority of executives are known to read only the *introduction and conclusions* of reports. For this very reason the concept of the Executive Summary has evolved. Technical readers often want to read selectively as well. They might want to skip the background section, page quickly through the material on equipment and procedures, and settle in to read the discussion of results. Whatever segments your reader wants to read, it is your task as a writer to make it easy to find these parts.

The easiest and most effective way to guide the reader through the text is using headings and other graphical signposts that forecast the content of various parts of the manuscript and draw attention to what is considered important. If the writer provides headings and signposts that clarify the organization of the document, the process of reading can speed up. In addition, a document with plenty of headings and other signposts looks attractive and inviting and reflects planning and care on the part of the writer; it therefore establishes in the reader a positive attitude toward the document and helps to make it more persuasive.

Effective Headings

Headings can help the reader in three ways: they are *verbal clues* in that they provide initial understanding; they are *visual signposts* in that they provide quick reference to parts of the document; and they are *physical markers,* making it easier to handle a document. Effective headings stem from good outlining techniques and precise phrasing.

Although headings help the reader by providing verbal clues and visual breaks, their effectiveness depends to a high degree on how informative they are. Effective headings clearly forecast the content of the section to follow, allowing the reader to decide whether to read without fear that he'll miss important information. This means that even when choosing headings for the major parts of a technical report, the effective writer goes beyond generic headings, such as APPROACH, ANALYSIS, or RESULTS, and makes the headings reflect the substance of the writing. The effective writer uses *headlines* rather than headings, for instance, ADVANTAGES OF INTRODUCING COMPUTER-OPERATED CUTTING MACHINES rather than COMPUTER-OPERATED CUTTING MACHINES. Effective headlines are especially useful in memos, the standby vehicle of communication within an organization. As discussed earlier in this chapter, a good subject line does more than indicate the subject discussed; it focuses on the purpose of the message and thus emphasizes its most important aspect.

Use of White Space and Graphic Embellishment

The visual appearance of the printed page is the first impression of any written message; like the grooming and dress of a person, it affects our attitude, and as with people, the first impression is often crucial. It can help us to get the attention of the reader and to make reading an efficient process. Taking care to make your writing look attractive, even if it is just a memo, pays off in the long run. Your reader might tend to pick your report first, before reading the rest of the pile, just because he knows it's easy to read.

One of the simplest devices for creating attractive layouts is *white space,* that is, the part of the page that is left entirely blank. Plenty of white space on a page gives the eye a break. White space between paragraphs makes the logical breaks in the texts also physical breaks. White space around headings focuses attention on the headings themselves. Plenty of white space in the margins not only provides room for notes, but also narrows the width of columns and thereby increases speed of reading. Plenty of white space around figures provides the emphasis they are meant to have.

Similarly, other graphic embellishments help the reader: *bullets* and *lists* break up the monotonous look of a page. Larger, *varied type sizes* add interest to the message. *Shorter paragraphs* and use of *italics,* **boldfacing,** CAPITALS, and other typographic devices make your documents more accessible.

Throughout this chapter we have discussed guidelines that help to organize technical documents. One of the advantages of structuring is that good structure can be repeated; the generic outlines presented here can be applied and modified over and over again. Moving from general to specific, using internal forecasts and summaries, using parallel structure, applying the basic patterns of development, and making structure apparent through headings and other signposts are all strategies that help you save time and effort.

REFERENCES AND FURTHER READING

Brunner, I., J. C. Mathes, and D. W. Stevenson. 1980. *The Technician as Writer.* Indianapolis, IN: Bobbs-Merrill Co., Inc.

Carosso, R. B. 1986. *Technical Communication.* Belmont, CA: Wadsworth Publishing Company, Inc.

Drucker, P. 1986. *The Practice of Management.* New York: Harper & Row, Publishers, Inc.

Weiss, E. H. 1982. *The Writing System for Engineers and Scientists.* Englewood Cliffs, NJ: Prentice-Hall, Inc.

Chapter 3

Writing and Revising the Rough Draft

The ineffectiveness of their own writing perplexes many engineers and scientists. Their peers and supervisors tell them that their memos and reports are "fuzzy" and "hard to follow." In response to these vague complaints, these troubled writers often resort to laboring over every word, phrase, and paragraph as they draft their reports and memos. All at one time, they try too hard to follow all the rules their English teachers taught, to search for the magic words that will produce understanding in their readers, and to rearrange those magic words into flawless prose.

While these ineffective writers' diligence is commendable, they should know that they are going about writing the hard way. The added work at the draft stage may eventually lead to frustration and loss of interest in writing when that laboriously written draft comes back for extensive revisions. The negative feedback loop comes full circle when the lack of interest in writing produces even fuzzier, harder-to-follow writing and an even more frustrated professional.

The solution to most writing problems, as this chapter explains, lies not in laboring over words and phrases at the draft stage, or in searching for the magic words (which probably don't exist), but in spending more time in the stages of the writing process which come before and after writing the rough draft: organizing the information (see Chapter 2) and revising the rough draft.

This chapter makes a few comments about writing the rough draft, but it dwells mainly on the critical stage of revising that first draft. It explains how to revise to avoid the charges of fuzzy and hard to follow. Finally, it discusses editing, or what is known under other circumstances as troubleshooting—the systematic tracing of symptoms to their underlying causes.

GENERATING A ROUGH DRAFT

We would say more about the actual process of writing a draft if the mysterious creative processes that constitute thinking were better understood. We can't any more explain what happens inside a writer's head when he or she starts to generate a memo or report than we can explain the process by which a potter throws a pot.

What we can say is that, besides preparation and organization, a writer needs discipline—not inspiration—to get that draft done. It takes discipline to write in the face of uncertainty (for no one can know all the relevant facts). It takes a further measure of discipline to deal with the writer's nemesis: the deadline. Everything written could conceivably have been made better if the writer had had just a bit more time—but time is something engineers and scientists never have enough of, and decisions have to be made whether or not the critical memo or report is ready.

Beyond that call for courage, our advice on writing drafts is to stick to the following cycle for getting words on paper efficiently so the more important business of revision can begin:

1) Start by expanding on the easiest part of the outline. Complete whole sections at a time, but take a break about every hour so that you can maintain your efficiency.
2) Write as quickly as possible; if you are familiar with the material, you should be able to write 500–700 words per hour. Do not stop to check spelling, browse through a thesaurus, or ponder punctuation. If you are missing a few numbers but otherwise know what you want to say, keep going and look up the facts later. Put all of your mental energy into developing paragraphs from each topic in the outline.
3) After you finish a section, let it sit overnight.
4) After this cooling-off period, revise the section according to the advice in this chapter.
5) Give the draft to one or more colleagues for comments. Choose readers whose judgment you trust and whose suggestions will be fair and firm. In addition, as Chapter 1 suggests, give portions of your draft as they develop to those who must approve the final document. Above all, have your draft reviewed by someone who is not familiar with the project on which you are writing so that any jargon or shoptalk

can be exposed for what it is—an avoidable rhetorical short-cut.

6) After a second cooling-off period, revise again, this time giving particular attention to consistency and emphasis.

Admittedly, some successful writers do very well by putting most of their effort into the draft stage. Zane Grey, for example, is said to have dictated his novels at one sitting, without benefit of an outline or other planning. He rarely had to make revisions. But unless you have an extraordinary gift, you would do best by following the six steps listed above.

THE IMPORTANCE OF REVISION

If efficiency is really an issue, why is revision so important? Why not write correctly the first time? The simple answer is that it can rarely be done. The justly famous *Elements of Style* cautions writers: "When you say something, make sure you have said it. Your chances of having said it are only fair."

To that caution add this advice: you must say something before you worry about whether you have said it well; and to worry about saying something and to worry about saying it well are too much to worry about at once. One of the characters in Albert Camus's *The Plague* illustrates these points by displaying all the typical frustrations of the misguided perfectionist's approach to writing a draft. This character takes pen in hand, and without first determining purpose, audience, outcome, or doing any other planning, writes something akin to "The rider rode over the hill on a fine horse." Throughout the novel, Camus describes the character's frustrating attempts to attain perfection by fiddling with syntax and word choice. The character's novel never progresses past the first sentence.

Another illustration of the same problem—worrying about too much at once—comes from considering the process of learning to play golf. Probably more has been written on the ideal golf swing than on any other subject in sports. Every aspect of the golfer's stance, grip, and swing has been analyzed. Sam Snead, whose own swing is so smooth and effortless that it defies analysis, comments that while it is fine to be a student of the game, a golfer can only think of a few things at a time when actually golfing. When a beginning golfer tries to remember every piece of advice during a swing, the results are disastrous. As Mr. Snead points out, we'd have the same bad luck if we pondered every detail of our eating green beans—we probably wouldn't be able to hit our mouths with a fork. The extension of that advice to writing is obvious: don't try to think of everything at once, or you will never get anywhere.

A final reason for why we should draft a report swiftly but revise it slowly and carefully comes from the appreciation that composing is, after all, a complex process. It consists of at least two separate mechanisms. One produces diverse possibilities. The other, something akin to natural selection, chooses the best of the lot. This second mechanism in the composition process, which we will refer to as revision, is discussed at length below because it is crucial to excellence in writing.

THE GOAL OF REVISING

As Chapter 1 points out, you must have a clear purpose in mind before you begin any writing work, and so it is with revising. Your goal in revising any draft should be *clarity*; that is, to convey your message transparently. Key attributes that bring about that transparency are concise expression and sound logic.

Because that is advice you have heard before, we worry that perhaps you won't take it seriously. Therefore, we stress that like other admonitions such as "Wear seat belts" and "Eat your vegetables," "Write clearly" is important enough that it bears repeating. Unlike these other bits of wisdom, which require only presence of mind to follow, "Write clearly" requires some coaching—provided below—before the advice can be followed to advantage.

CHOOSING THE RIGHT WORDS

As with other skilled activities, writing requires correct attitudes as well as knowledge. One attitude that is key to successful revision is the attitude that individual words must be chosen to achieve clarity.

How else do we choose words? Doesn't everyone at least try to write clearly? Let us look at some examples that show that writers often choose words with something other than clarity in mind.

Choosing Words Out of Habit

In his essay "Politics and the English Language," George Orwell chews up modern writing and spits it out with disgust. His thesis is that writing is often bad because we choose the overworked word or phrase rather than struggle to find the clearest expression:

> As I have tried to show, modern writing at its worst does not consist in picking out words for the sake of their meaning and inventing images in order to make the meaning clearer. It consists in gumming together long strips of words which have already been set in order by someone else, and making the results presentable by sheer humbug. The attraction of this way of writing is that it is easy. It is easier—even quicker, once you have the habit—to say *In my opinion it is not an unjustifiable assumption that* than to say *I think*.

The engineer or scientist who sends out a letter beginning with "I would like to take this opportunity to say that . . . " has chosen the most convenient words (those that come to mind first), not the clearest words. With the correct attitude, he or she can revise to fix the problem.

Choosing the Best-Sounding Words

Occasionally we are all guilty of choosing words solely to embellish a poor idea or unpopular statement. We choose our words carefully, but not with clarity as our goal. Instead we hope to sugar-coat the unpleasant with a layer of euphemism.

Certain professionals, many of whom work for the government, have become experts at improving appearances by using words that disguise reality. Consider the names of our least favorite government agencies: **Internal Revenue** is far more pleasant sounding than the word **taxes** which it in fact collects, just as **Selective Service** seems more agreeable than the **draft** it administers. In neither case does the government's choice of words change in the least what the agency does.

The art of choosing words that disguise rather than reveal is practiced outside of government as well. Doctors work hard at using obscure Latin-based words rather than familiar words that convey the unpleasant or the unhealthy: they might choose the word

diaphoresis instead of the simple, clear **sweat,** and they would for certain choose **emesis** over **vomit.** Real-estate people often use **cozy** to describe a house that is much too small, and **gracious Victorian** to describe a house that is much too large, old, and drafty to be appealing. Of course, all of these substitutions detract from clarity while doing nothing to the facts.

Choosing Words That Impress

Simply put, sometimes misguided writers choose words to impress rather than inform. We who work at a university see this problem in our students; we also see where the problem originates: in our colleagues' writing. Now, a large working vocabulary is a great thing for an engineer or scientist to have. It shows attention to words and experience with words that can only have come through extensive, careful reading. It allows proper discrimination of words whose meanings vary in subtle ways—**advance** vs. **advancement** and **historic** vs. **historical**, for example. But it is not so valuable that it has to be exercised in every memo and report. Writers who, like high school students, choose unusual, even obscure words to express familiar concepts do not serve their readers' interests. In our experience, good ideas and clear recommendations impress readers; big words and pretentious phrases irritate them. Consider this particularly annoying example:

> As a result of the fact that four of our members could not be present at our last scheduled meeting time, the liberty was taken to cancel that assemblage. It is, therefore, planned to consider the same agenda items at the newly scheduled aggregation. Therefore, the conference will be convened in accordance with the same agenda, repeated herein for convenience.

This memo gave its author a chance to show off, but it didn't do much for the reader, who would appreciate a version revised to enhance clarity. Later on, we will consider a clearer, less irritating version of this memo.

As an aside, notice that the above example suffers from not only long words, but also from a long-winded style. The two problems—like swamps and malaria—almost always come together. Having dealt with one of the problems, we next discuss the other: wordiness and how to combat it.

ACHIEVING A CONCISE STYLE

Editors of scientific journals must frequently send back articles for revision that are longer than they are informative. The authors' response is sincere and predictable: "Every word is vital." The editors know better. In fact, since these editors know how to revise, they could do the shortening themselves if they had the time.

Indeed, the authors' lack of time is probably what caused the article to run long to start with. As Blaise Pascal said, "I have made this letter longer than usual because I lack the time to make it shorter."

To achieve brevity in all your writing, you must make an effort to look for the same problems that editors of scientific journals look for when they revise:

- empty words and phrases
- passive voice verbs
- expletive constructions

Eliminating Empty Words

Different writing books use different terms for the words that don't carry their own informational weight: *low-information-content words, deadwood,* and *wooly words* are three terms that come to mind. All three point to the fact that some words aren't essentially necessary.

Or put concisely, all three mean unneeded words.

Sometimes a word is simply *redundant*; that is, it gives information that is already given. In the following example, redundant words are crossed out:

Some Asian countries have caused ~~hostile~~ antagonism because of their policies to eradicate ~~completely~~ competition from ~~foreign~~ imports.

Antagonism implies hostility; **eradicate** means "to get rid of completely"; and **foreign** is the only sort of imports one can mean. The extra words add not emphasis but only the impression that you are not thinking carefully about what you are saying. Despite the illogic of redundant expressions, some have crept into common use. Perhaps logic will win out eventually, and expressions like **hot** water heater (it really heats *cold* water), **free** gift (if it isn't *free*, it isn't a *gift*), and tuna **fish** (as in *trout* fish?) won't be around.

Other times we add words at the beginning of a sentence as a way of getting up to speed. For example, we might say *as a result of the fact that* instead of *because* when we aren't sure what caused what. Your reader, who is already up to speed, would appreciate information instead. Therefore, when revising, look especially critically at the beginnings of sentences:

> **Basically**, the mechanism suffers from three fundamental design flaws. (As readers, we get the impression that, basically, the author hasn't yet thought of what those flaws might be.)

> **Using a large number of very complex experiments**, we will determine the optimum configuration. (This sentence, written in the familiar style of the proposal writer who sees a problem but no answer, never does say anything. What he or she might mean is, "We're looking to change lead into gold, and we want funding so that we can buy a sample of each.")

Still other times we lapse into using a verb and a noun when another verb would take the place of both. **To take into consideration** really means "to consider," just as **to give assistance to** really means "to assist" (or even better, "to help"). The wordiness in the example in the left-hand column stems from weak noun-verb combinations. The concise sentence in the right-hand column relies on forceful verbs.

Weak:	*Strong:*
To **make a reservation** for a table, **give notification** to the chairman of the number of guests that you **have plans** to bring.	To **reserve** a table, **notify** the chairman of the number of guests that you **plan** to bring.

Avoiding Passive Voice Verbs

Most modern guides to writing style, including our favorite, *The Elements of Style*, recommend using verbs in the active voice whenever possible. The only problem with that advice, we've found, is that not everyone is exactly sure what makes a verb active. So that you can distinguish between passive voice verbs, which you should use sparingly, and active voice verbs, which you should use often, we will take a brief trip through the Valley of the Shadow of Grammar. Heed the advice but skip the following section if you already are aware of that distinction.

In any grammatically correct sentence, each verb has a subject, but not all verbs convey the action of their subject. Sometimes the subject of a verb is being acted upon (passive) rather than active. Consider these sentences:

1) The cat ate the rat.
2) The rat was eaten by the cat.

The subject of sentence 1 is **cat**. The cat is actively doing the eating. The subject of sentence two is **rat**. It is not doing anything (certainly it isn't eating).

These two sentences convey the same meaning, yet they differ in their effect on a reader. In sentence 1, the action goes in the same direction the reader scans. In sentence 2, the action works backward, from the end back toward the beginning. A reader detects a difference in emphasis as well. In sentence 1 the emphasis is on the *cat*, while in sentence 2 it is on the *rat*.

Whether to use active or passive voice is a choice we must make whenever we write most sentences. (Note, however, that not every verb can be written in both active and passive voice: only verbs like **to kill** and **to bother** that show action to something or somebody have two voices; verbs like **to be** and **to die** don't.) We have already said that engineers and scientists should, as a rule, use active voice. It has these advantages:

- **It is more natural.** (Can you imagine hearing "A cracker is wanted by Polly"?)
- **It is more informative.** (Active voice makes clear who is doing what. Passive voice many times does not: "The funds were diverted illegally." Who is the culprit? The reader doesn't know.)
- **It is shorter.** (Passive voice always requires more words than active voice: "A cracker is wanted by Polly" vs. "Polly wants a cracker.")
- **It is more compelling.** (If you want something done, use active voice. A sign stating "The last person to leave must lock all doors

and windows" will improve security; one stating "All doors and windows must be locked at the end of the day" will not.)

What does passive voice do well? It emphasizes *what is being acted
upon*. Sometimes that is preferable: **The Interstate Highway
System will be completed in 1989,** a passive voice sentence,
works well because what is being completed is much more important
than who is doing the completing. Notice that this chapter section on
the importance of active voice does indeed contain a few passive
voice verbs. Can you find them? Can you see how they emphasize
the topic under discussion?

If you can answer those questions, you probably can agree with us
that what the style guides want writers to avoid is not the use of
passive voice, but its overuse. If you choose verb voice carefully to
achieve brevity and proper emphasis, you will improve your writing
greatly, which is what the style guides want.

Avoiding Expletive Constructions

Sometimes we need to communicate a simple, unadorned idea.
That requires adding words to make a complete sentence around that
idea. Grammarians call the grammatical filler that we add in such
cases *expletive constructions*, or simply *expletives*. To turn the idea
"raining" into a sentence requires the expletive **it is**; the sentence **It
is raining** results.

It is and **there are** are the most common expletive constructions.
They are handy devices; "It is raining" is certainly a more idiomatic
and natural response to the question **What's the weather like?**
than is "Rain falls." However, you should look for ways to eliminate
expletives from your rough draft because they make a literally mean-
ingless word the subject of a sentence. Our advice is to use a "real"
subject instead. As the following examples suggest, such revisions
are usually easy:

Original:	*Revision:*
There are three deer that live in those woods	Three deer live in those woods.

It is imperative that new employees file a data card with the Employment Office.	New employees must file a data card with the Employment Office.

In our experience, engineers and scientists who use many passive voice verbs also overuse expletives. Probably they use both strategies to avoid use of personal pronouns—especially the dreaded I. Indeed, some companies' style manuals, in an effort to promote a veneer of objectivity, discourage personal pronouns.

We think that policy is old-fashioned. More modern style guides, such as the *CBE Style Manual* (Council of Biology Editors), encourage writers to use personal pronouns. Whether the phrasing is **It is recommended that** or **The recommendation is made that**, someone is behind that recommendation: the phrasing **I recommend**, besides being shorter and more direct, adds, along with extra information, a sense of conviction.

Achieving a Concise Style: An Example

With the above specific strategies for achieving brevity in mind, let us look again at the irritatingly wordy memo about the meeting. It is repeated below, first with wordy phrases in boldface, and then in a revised form that is much more concise but just as informative:

Wordy:

As a result of the fact that four **of our** members **could not be present** at our **last scheduled meeting time**, the **liberty was taken to cancel that assemblage. It is, therefore, planned to consider the same agenda items** at the **newly scheduled aggregation, which will convene Friday.** Therefore, the conference **will be convened in accordance with the same agenda, repeated herein for convenience.**

Concise:

Because four members **couldn't be** at the last meeting, I **rescheduled** it. **We will** consider the same agenda (**see the attached copy**) on Friday.

The original 64 words boil down to 24. Admittedly, the example represents extreme wordiness, but using the same strategies will allow you to shorten—and thus improve—almost anything you write.

ACHIEVING LOGICAL WORDING

Good writers are to readers as good guides are to travelers. A writer has a goal (a recommendation, a point of view, or a request) that he or she wants a reader to reach, just as a guide has a destination to which he or she takes travelers. Good writers carefully lead readers toward that goal; ineffective writers merely point and leave the reader to find the path to that goal.

Writers whose work is labeled "hard to follow" don't put enough of their thinking on paper, or they put it down in a form that is not logical. That is, their writing does not follow, either on a large or small scale, the pattern that a reader would expect.

To a great degree, the cure for hard-to-follow writing is better organization (Chapter 2). But careful writers can do much at the revising stage to improve control of readers' expectations by writing better sentences. By better, we specifically mean sentences that better show *what is important, what is equally important,* and *what is less important.* We also mean sentences that make clear references so that the writer answers readers' questions—without posing new ones.

Showing What's Important

Readers determine what's important by looking at sentence structure. As a writer, you therefore must structure sentences to give the correct clues. Consider this sentence:

> Since it operated at a $150,000 deficit last year, the Training Department is investigating ways to cut instructional costs.

Which fact is more important? That the department lost money last year or that it is now investigating cost cuts? Which fact would we as readers expect to find out more about in the subsequent text?

Clearly, the sentence structure tells us that the investigating is the important fact, the one we expect to read more about. Let's look at the grammar of the sentence to see why. The sentence has two *clauses*, that is, two separate statements, each of which has a subject and verb. But they are two very different kinds of clauses. The first is called a **subordinate clause**; the word it begins with (**since**) is called a **subordinator; since** by its nature makes the clause it begins less important, or put another way, subordinate to another clause. The second is called an **independent clause**; it could stand by itself independently as a sentence. It is the clause to which the "since" clause is subordinated.

Proper subordination—the proper use of subordinate clauses for less important statements—is a key to easy-to-follow writing. Subordination means that facts, instead of being artlessly strung end to end, can be related, compared and explained. But what if you wish instead to show that two facts are *equally* important?

The flip side of subordination is **coordination**. The structure of the sentence below shows the two clauses to be equally important:

Lower energy costs have stimulated economic growth in the Midwest, but they have caused hardship in oil-producing areas like the Gulf Coast.

The key word in this sentence is **but**. It, along with **and** and **or**, are known as *coordinate conjunctions*. When one of these words joins two clauses, a reader will treat the two statements as equals. From the structure of the sentence on energy costs, one would expect to read more about both economic growth and hardships.

The key point is this: **Good writers use subordination frequently**. They use coordination infrequently. Inexperienced writers do just the opposite. They seldom subordinate facts, while they overuse the coordinate conjunction **and** in the manner of a grade school student relating an event: "Me and Sarah found some pop bottles and we took them to the store and the man gave us some money and we bought some candy." So, emulate good writers by using subordination liberally. It will make your writing easier to follow because the important information will stand out.

Making Clear References

Good communication entails making sure that each party understands what the other is talking about. When you revise your writing, you can improve it by making sure that on the sentence level your reader always knows what you are referring to.

Problems to look for are of two kinds. First, *pronouns* (those words like **it** that we use to stand in for *nouns* like **Yosemite National Park**) have to point clearly to the word or words for which they stand. Second, words or phrases that describe (modify) something in a sentence must be positioned so that no reader can misread what is being described. Engineers and scientists, in the interest of the precision that is their hallmark, should take special care to avoid these vague pronouns and misplaced modifiers.

Below are examples of both kinds of errors and suggested revisions:

Vague pronoun:	*Revision:*
Putting marigolds next to cucumbers will keep the rabbits away from **them**. (Vague pronoun. Will rabbits stay away from the *cucumbers*, the *marigolds*, or *both?*)	Putting marigolds next to cucumbers will keep rabbits from eating the cucumbers.
The task force failed to complete its study of the mine accident. **This** was the subject of a scathing editorial in the union newsletter. (Vague pronoun. What singular noun does **this** stand for?)	The task force failed to complete its study of the mine accident. This failure was the subject of a scathing editorial in the union newsletter.

Misplaced modifier:	*Revision:*
Upon completion of the installation, the contractor shall leave the premises **in an orderly condition**. (Misplaced modifier. Does "in an orderly condition" describe the *premises* or *manner in which the contractor shall leave?*	Upon completion of the installation, the contractor must ensure that the premises are in an orderly condition.
Being of sound mind and body, the will was certified by my aunt's lawyer. (Misplaced modifier. The introductory phrase is what is known as a *participial phrase*, **being** being a *participle*.	Because my aunt was sound of mind and body, the attorney certified the will.

Such phrases must describe nouns; in this sentence the phrase is next to the noun **will**. Yet what the phrase describes is **my aunt**, a noun that isn't even in the sentence. This particular modification problem is called a *dangling participle.*)

Hoping for a better crop, the field was irrigated frequently. (Another dangling participle. The field can't *hope*; the farmer can, but he isn't in the sentence. Such unfortunate sentences create yet another argument in favor of active voice: using passive voice makes it all too easy to create dangling participles).

Hoping for a better crop, the farmer irrigated the field more frequently.

Following Writing Standards

Your English teacher badgered you about all those "rules" of writing for a reason. We think. He or she wanted to teach you the importance of correctness. Certainly, for an engineer or scientist to write carelessly gives an unfortunate impression of generally sloppy work habits.

However, we relegate discussion of those rules (to be enumerated shortly) to this short subsection of one section of one chapter (plus Appendix 1). We don't badger you about them. What we do is point out that following writing standards is just one more way to keep your reader with you. But it is only one of the things to think about when you revise.

We also point out that many of the rules of writing are really conventions—customs but not law. The English-speaking world has established them to standardize discourse. When you as a writer fail to follow one of them, a reader might be briefly distracted from what you are trying to say. So, if you choose to flout convention, know that your reader will notice—a fact that makes occasional unorthodoxy a useful device for emphasis. These are some of the conventions that we were taught as rules:

1) Never end a sentence with a preposition. (As in "Prepositions are not the sort of things to end sentences **with.**")
2) Never start a sentence with **and, or,** or **but.** (**But** we start sentences with coordinate conjunctions all the time!)
3) Never write a sentence fragment, that is, something punctuated as a sentence that lacks an independent clause. (**Well, almost never. Fragments have their place. Like in advertising.**)
4) Do not split infinitives; that is, do not put a modifier between the **to** and the rest of an infinitive (as in "**to boldly go** where no man has gone before").

We all know many good communicators who know about these conventions but don't always heed them. However, good writers *always* heed other things your English teacher taught you which really are rules. Disregarding them results in illogical writing at worst, and careless writing at best. Engineers and scientists don't want to be labeled either illogical or careless. Therefore, you should double-check during the revision process to make sure that you have followed these "real" rules closely:

1) Subjects and their verbs must agree in number; that is, if the subject is plural, then the verb must be as well.
2) Pronouns and their antecedents (that is, the words they stand for) must agree in number; for example, "this" should take the place of something singular, while "these" should take the place of something plural.

Adhering to standards of English, whether mere conventions or real rules, helps you communicate with your readers, and every engineer and scientist should become a student of those standards. All the same, we want to emphasize that good writing does not come just from following rules and conventions any more than good tennis results just from following the rules and etiquette of the game.

REVISING AS TROUBLESHOOTING: SYMPTOMS AND CURES

So far our advice on revising has been geared toward developing attitudes: choose words carefully, work for a concise style, try to be logical. We now wish to reinforce those concepts by warning you of specific words that often mean a specific sort of writing trouble. Just as black exhaust smoke tells a mechanic what problem lurks under the hood, so should certain words alert you to specific writing problems when you are revising.

Symptomatic Word: "AND"
Cure: Use Subordination

Tighten up your writing by using subordination: clauses that begin with words such as **because, since, also, if, when, although, which, that, who, even though,** and **so that.** We found the following passage, peppered with **ands**, in a local newspaper. To the right it appears smoothed into something much more readable with the help of subordination (and active voice).

Original:	*Revision:*
This auction is being held to settle the late Mrs. Truttmann Estate and this represents a well-known New Glarus area family and this auction must be seen to be appreciated.	This auction will settle the estate of the late Mrs. Truttmann, who was a member of a well-known New Glarus family. You must see the items for sale to appreciate their cleanliness and excellent condition.

Here's another example, the last paragraph of a cover letter for a job application:

Original:	*Revision:*
In summary, I am confident that my qualifications suit me for the position, and I wish to be considered. I enclose a resume, and I hope you will consider it carefully.	To summarize, because I am confident that my qualifications suit me for the position, I wish to be considered. Enclosed is a resume, which I hope you will consider carefully.

A final comment on **and**: besides checking each **and** to see if subordination would work better, you have to check for *parallel structure.* That is, whatever is joined in a sentence by an **and** (or **but** or **or**) must grammatically balance. This concept is simply an extension to the sentence level of Chapter 2's discussion of parallel structure in organizing writing.

Here are examples of **and** signalling problems with parallel structure:

Problem:	*Revision*:
You can put a hole in steel by drilling, punching, and a cutting torch will burn a hole. (**Drilling** and **punching** don't balance **a cutting torch will burn a hole**, which is a complete clause.)	You can put a hole in steel by drilling, punching or burning.
The fracture zone can be increased by making some of the shot holes longer than where breakage occurs, and then load and fire the holes as usual. (Not easy to follow! The main problem is that **load** and **fire** should match **making** to achieve parallel structure.)	You can increase the fracture zone by extending some of the shot holes beyond where breakage has occurred, and then loading and firing the holes as usual.

Symptomatic Word: "VERY"
Cure: Eliminate Imprecise Words

The language of engineers and scientists should match other aspects of their work in its tangible, empirical and verifiable nature. That language cannot then include words, chiefly adjectives and adverbs, that make unverifiable, imprecise statements. **Very** is such a word. To say something is **cold** is fine, although we would be happier with a quantified description; to say that something is **very** cold is not fine because the added word does nothing for the reader.

Proposal writers must, when revising, be especially vigilant for imprecise modifiers: **important, complicated, high-quality, revolutionary**, or even worse, **very revolutionary**. The tone of proposals must be upbeat and positive, but engineers and scientists should at all times leave the hyperbole (exaggeration) to the pros on Madison Avenue. We were cured of using **important** by a department store ad highlighting polo shirts in "a choice of sixteen important colors." Here is an example of a student engineer's writing that cries out for revision because of the frequency of imprecise modifiers:

Original:	*Revision*:
Having a liquid nitrogen facility here would be very beneficial to this great university. The university would be able to supply helium to a lot of research groups at a very reasonable price. The university would also make large sums of money on the sale of this important product, enough to pay back the investment within a very short span of time.	The university would benefit from having its own liquid nitrogen facility. Because of the interest in cryogenics and superconducting materials, seven different research teams now use on the average a total of 600 liters of liquid nitrogen per month. My calculations show that, given current prices and interest rates, the university could recoup its investment in a liquid nitrogen generating facility in 3.5 years.

Symptomatic Word: "THIS"
Cure: Make Clear References

In his book *The Careful Writer,* Theodore Bernstein warns us all of overusing the word **this**:

> In recent years there has been a lava-like spread of the word *this*: "This is murder"; "See that bearded guy across the room? Well, this is a real painter." . . . *This*, which has always referred to something present or near, is rapidly supplanting *it* and *that* even where no proximity is indicated. This [sic] is linguistic progress?

What Mr. Bernstein objects to is not the use of the handy pronoun **this**, but rather its careless, habitual use. We have witnessed the lava-like spread of **this** that Bernstein mentions. In our experience it has occurred almost totally in bad writing. We have become particularly sensitive to **this** when used as the generic, universal way to start the second sentence of any paragraph, as in these examples:

> The present methods of storage used in the warehouse are not capable of providing space for all incoming shipments. **This** results in the need to secure off-site storage space.

In past years, no organizational planning has been conducted when staff or equipment has been added. **This** has created an unorganized, inefficient facility layout.

As we have already discussed, pronouns must have a clear reference: a recently stated noun or noun clause for which the pronoun stands. Now, in the examples above, one can guess what the writer means by **this**. In the first example, he or she means "inefficiency of the present storage methods"; in the second, he or she means "the lack of organizational planning." But good writers do not make readers do the writers' work, which is to clarify facts and relationships.

Therefore, heed this advice for containing the spread of **this** and the fuzzy writing it creates. Every time you come across **this** in a rough draft, make these checks:

1) Does a clear reference exist?

Original:	*Revision*:
The problems with the current layout which were previously mentioned will be eliminated. **This** will result in smooth production flow and increased efficiency.	The previously mentioned problems with the current layout will be eliminated. Smooth production flow and thus increased efficiency will result.

2) Would **it** or **they** work better?

Original:	*Revision*:
The main gear has three damaged teeth. **This** causes reduced transmission efficiency.	The main gear has three damaged teeth. They cause reduced transmission efficiency.

3) Would the statement beginning with **this** work better as a subordinate clause connected to the previous sentence?

Original:	*Revision*:
Entry-level engineers sometimes struggle the first few months on the job. **This** is because they are	Entry-level engineers sometimes struggle the first few months on the job because they are used to

| used to working on well-defined problems. | working on well-defined problems. |

Making these checks will help you rid your writing of the most easily avoided sort of fuzziness: vague pronouns.

Symptomatic Word: "UTILIZE"
Cure: Use Ordinary Words

To us, **utilize** is a fat, ugly, pretentious word. None of your authors has used it in the last five years—not even once. Nevertheless, it is a word that our students use frequently. Just as frequently we cross it out and substitute the short, simple word that means the same thing: **use**.

Why does **utilize** so frequently appear in student writing? Because professors use that word, and many other words like it, that induce a hypnotizing polysyllabic cadence toward which one gravitates when rhapsodizing about one's area of expertise (in other words, **utilize** goes great with pontification). So hypnotizing is that cadence that these professors even manage to stick extra syllables into and onto words: **pronounce** becomes **pronunciate; orient** becomes **orientate**; and **regardless** becomes (excuse us while we gag) **irregardless**. As engineers and scientists, our job is not to rhapsodize or hypnotize but rather to inform: to explain, to persuade, to relate, to describe. As we have reminded you often in this chapter, those tasks require a simple, ordinary vocabulary. When revising, you should, almost as a reflex, cross out pretentious words like **utilize** and replace them with their blue-collar counterparts. The following list shows how easy those changes can be:

Fat, ugly, pretentious:	*Better:*
Demonstrate	Show
Endeavor	Try
In close proximity to	Close to, near
Facilitate	Help
Usage	Use
At that point in time	Then
Has the capability to	Can
Discontinue	Stop

Look further at the sort of company that **utilize** keeps in the following passages. Also study the suggested revisions.

Pretentious:	*Clear*:
Basically, it would not be unreasonable to assume that in the foreseeable future our very unique but nonetheless inefficient hand-fabricated toy production division will have to be terminated.	We may soon have to sell our hand-made toy production division because of its inefficiency.
During the course of his education, it is absolutely essential that an undergraduate student secure the acquisition of the basic fundamentals of engineering.	An undergraduate engineering student must learn the basics of engineering.

REVISING: A SUMMARY

What these passages and revisions show is that **utilize** and other pretentious words often pop up next to other signs of bad writing that we have discussed: expletives, passive voice and redundant expressions. All these unfortunate word choices occur because writers lapse into using bad writing strategies: writing to impress rather than inform; writing out of habit; and making the reader do part of the writer's job. Remember, we want you to worry about none of these problems while you are writing first drafts; but we want you to worry about all of them in turn as you revise.

We hope that this chapter has given you strategies that will allow you to produce more effective writing. Certainly, you now know that revising a draft is much more than checking spelling and adding a few commas here and there.

If some of the advice on revising sounds familiar, it is because it hasn't changed much. Neither has bad writing changed much. Mark Twain gave much of the same advice in his 1895 satirical essay "Fenimore Cooper's Literary Offenses." His comments, themselves a model of clarity, echo our advice on revising:

1. Say what [you] propose to say, not merely come near it.
2. Use the right word, not its second cousin.
3. Eschew surplusage.*
4. Avoid slovenliness of form.
5. Use good grammar.
6. Employ a simple and straightforward style.

But do all of these things as you revise, not as you try to write a first draft.

REFERENCES AND FURTHER READING

Bernstein, Theodore. 1965. *The Careful Writer*. New York: Atheneum Publishers.
Camus, Albert. 1969. *The Plague*. Translated by Stuart Gilbert. New York: Alfred A. Knopf, Inc.
CBE Style Manual Committee. 1983. *CBE Style Manual: A Guide for Authors, Editors, and Publishers in the Biological Sciences*. 5th ed. rev. and expanded. Bethesda, MD: Council of Biology Editors, Inc.
The Chicago Manual of Style. 1982. 13th ed. Chicago: University of Chicago Press.
King, Lester. 1978. *Why Not Say It Clearly?* Boston: Little, Brown & Company.
Orwell, George. 1968. Politics and the English language. In *Collected Essays, Journalism, and Letters, Vol. 4: In Front of Your Nose, 1945–1950*, Sonia Orwell and Ian Angus, Eds. New York: Harcourt Brace Jovanovich, Inc.
Robinson, Patricia. 1985. *Fundamentals of Technical Writing*. Boston: Houghton Mifflin Company.
Strunk, William, and E. B. White. 1979. *The Elements of Style*. 3rd ed. New York: Macmillan Publishing Co.
Twain, Mark. 1976. Fenimore Cooper's literary offenses. In *The Unabridged Mark Twain*, Lawrence Teacher, Ed. Philadelphia: Running Press.
Zinsser, William. 1980. *On Writing Well: An Informal Guide to Writing Nonfiction*. New York: Harper & Row Publishers, Inc.

*Remember, this is satire.

Chapter 4

Visual Displays in Technical Writing

When preparing reports, memos, or other kinds of writing, most engineers concentrate on the writing process itself. They spend most of their time writing and revising and revising again. Often only in the last stages of this process do they think of including visuals or exhibits. As a result, these visuals appear to be added to the text rather than an integral part of it.

Why do writers tend to avoid using visuals? Perhaps they feel visuals are difficult to prepare, require valuable space, and are expensive to produce. Perhaps being preoccupied with words prevents writers from thinking and writing in pictures.

On the other hand, why do certain publications seem to revolve around exhibits? For example, articles in the *National Geographic* are designed around photographs. And advertisements focus on graphical displays. In these cases visuals are the bait, the attention getter, and meant to entice the viewer to read the text. In fact, the *National Geographic* owes much of its enduring success to its pictures. With well-chosen graphical designs you can have a similar effect on the reader of your technical material. While scanning through a report he will probably pause for the exhibits. And if these are interesting and attractive, he is inclined to look at the document in more detail.

This chapter explores the communication value of visuals and the situations when engineers do well to use them. It also discusses the most common types of graphical presentations in technical writing and presents guidelines for producing accurate and attractive visuals and for integrating them into the text.

WHEN TO USE A VISUAL

Before answering the question of when to use visuals, we will discuss briefly how visuals function in the communication task. While the reader perceives words of text as a temporal sequence that he can follow one step after the other, he looks at graphical displays as spatial entities that he can decode at once. A visual in the text is a bit like a billboard along the highway: if properly designed, its core idea is understood at a glance. Of course, quick comprehension is desirable in most communication tasks.

In addition, visuals allow the writer to condense a great deal of information into compact form. Rather than wasting space, visuals in many cases save space. For example, lengthy explanations of tectonic movements of the earth crust are condensed into a simple diagram, or sequences of enrollment figures for engineering students are conveniently summarized in a table.

Further, well-designed visuals appeal to the reader's sense of aesthetics. They not only organize the flow of information from the graphic to the reader by making important things stand out, but simplicity, clarity, and interest in design help to make visuals attractive and hold the viewer's attention.

In addition, visual presentation in conjunction with verbal documentation appeals to different learning styles. Research of the learning process has clearly shown that comprehension follows both verbal and visual patterns. Thus, for most readers visuals supplementing the text make it easier to comprehend your message.

Finally, telling it twice using different techniques—showing in addition to telling—ensures that your audience can more easily recall your message. People tend to remember only 10% of what they have read, while they tend to remember 30% of what they have seen. By saying it both in words and in a picture, your audience may remember as much as 50% of your message.

Because graphical displays have these highly desirable qualities, developing an acuity in being able to show is almost as important for a writer as being able to describe in words. Yet, the countless variety of graphical forms does not lend itself to a simple set of rules that a writer can apply. In the following we offer only guidelines, being aware that there are many exceptions to these guidelines.

The most obvious situation for using a visual comes up when words won't do. For example, when you have a *picture* in mind while describing in *words*—the way to the bus stop or the arrangement of washers in a water faucet—or when you use your *hands* while describing *orally*, in such cases you would readily agree that a visual will do the job more efficiently. In other cases, you might have to record vast amounts of data, and a table or a line graph would do that clearly more efficiently than numbers interspersed in the text.

Also, in many situations words won't do as well as visuals. You should use visuals when information is faster and easier to understand in graphical form. A drawing, for example, illustrates at a glance how to insert a floppy disk into a disk drive; a simple bar graph clarifies the temporal relationships of the various phases of a project; a pie graph readily reveals the general distribution of expenditures in a company; a line graph demonstrates at a glance the linear relationship between atmospheric temperature and pressure. In fact, as soon as you list more than four or five numbers in a paragraph you should begin to consider if a visual would not do a better job in communicating these numbers.

Finally, you can use visuals effectively in order to highlight an important point. Even a simple exhibit consisting only of words—like the list shown below—will have this effect. Here white space, that is, space on the page that is left blank, different type styles, bullets, and positioning are the graphical elements.

Use visuals in these situations:
- when words won't do
- when information is faster and easier to understand in graphic form
- when a visual highlights an important point

CHOOSING THE RIGHT VISUAL

The many graphical forms available to the engineer can be used to communicate different purposes and various kinds of information and to appeal to different audiences. *Tables and graphs* summarize data and statistics, such as the steel production in the United States or the growth of the national debt; *charts and drawings* illustrate concepts, processes, or concrete objects, such as the greenhouse effect, the manufacturing process of car bodies, or the floor plan of an office building. In addition, some graphical forms are more difficult to read or interpret than others and should be reserved for expert or technical audiences. Other types of visuals are preferred by a general audience. In the following we outline the main functions and features of commonly used visuals and provide guidelines for their construction.

Tables

Tables are used to present large amounts of data and to give absolute values where precision is important. Tables *show well* many discrete data in a small space. However, tables do not show well trends or direction in data.

The system of rows and columns in a table makes it possible to group data effectively and therefore to make them accessible to the reader. As Figure 1 shows, data to be compared are arranged in vertical columns rather than rows, and these columns rather than a grid system guide the reader as he or she scans the numbers. Also, the most important column by position (first or second column) displays the organizing principle of the table, in this case increasing complexity of agricultural techniques used in Africa. A table that does not indicate the principle for its arrangement is like a report with all its sections mixed up.

Maize grain yield (kg/ha) in tied-ridge systems in Burkina Faso. DAP, days after planting.

Ridge system	Yield (kg/ha)		
	Low fertilizer	High fertilizer	Mean
No earth up	1040	1480	1260
Earthing up* at 30 DAP	990	1470	1230
Earthing up at 30 DAP; ridge tied every other furrow	1840	2540	2190
Earthing up at 30 DAP: ridges tied	2040	3280	2660

*Earthing up means forming ridges and cross ties manually and weeding simultaneously.

Figure 1. Example of statistical data arranged in a table. (Reprinted by permission from R. Lal, 1987, "Managing the Soils of Sub-Saharan Africa," *Science* 236:1071 [29 May]. Copyright 1987 by the AAAS.)

Rather than only showing numerical data, a table can also display effectively verbal and graphical information. Figure 2, for example, summarizes the major thought processes taking place in the left and right hemispheres of the brain. In this case, the table not only saves space but also makes it easy to find certain information and to remember it.

Left-Hemisphere Processing	Right-Hemisphere Processing
Interested in component parts—detects features	Interested in wholes—integrates component parts and organizes them into a whole
Analytical	Relational, constructional, pattern-seeking
Sequential processing, serial processing	Simultaneous processing, processing in parallel
Temporal	Spatial
Verbal—encoding and decoding speech, mathematics, musical notation	Visuo-spatial, musical

Figure 2. Example of verbal information arranged as a table. (From L. V. Williams, 1983, *Teaching for the Two-Sided Mind,* Englewood Cliffs, NJ: Prentice-Hall. Copyright © 1983 by Linda Verlee Williams. Reprinted by permission of Simon & Schuster, Inc.)

GUIDELINES for constructing tables:

- Place columns to be compared next to each other.
- Make headings and data reflect an organizational principle (priority, descending order, alphabetical order).
- Label each column and row.
- Include units of measure in the headings.
- Align decimals in a column.
- Put table number and title at the top.
- Use footnotes for more extensive explanations of data or headings.

Graphs

Engineers use graphs to analyze and show relationships between two or more variables. In this overview we distinguish *line graphs, logarithmic graphs, bar graphs, pie graphs,* and *pictographs.* These graphs serve different purposes and are of varying complexity. Sources listed at the end of this chapter provide information on details and modifications of the basic graph forms.

Simple Line Graphs

Line graphs are used to show trends and relationships. They show well continuity and direction in one or more variables. The rectilinear coordinate system used in line graphs allows plotting values of a quantity as a function of another variable. Because the resulting line is easily interpreted in terms of increases and decreases, the line graph has become one of the most widely used graphs. Although the horizontal axis, or abscissa, most often depicts time, as in a graph of monthly productivity increases, line graphs are especially useful for identifying and displaying relationships between two variables, as shown in Figure 3. This graph, showing the radiation of a black body at different temperatures as a function of wavelength, illustrates that the scales of the horizontal and vertical axes must be chosen to ensure proper resolution of significant increases and decreases in the variables. In addition, while line graphs are an excellent graphical tool for comparing trends or changes of two or three variables, they become confusing if more than four lines that all move in different directions are shown. In such cases, separation into several graphs will solve the problem. Tick marks and tick mark labels should not interfere with the lines and are best placed outside the graph area.

GUIDELINES *for constructing line graphs:*

- Limit the number of lines on a graph to three or four.
- Distinguish different lines by design or color.
- Choose the range of tick marks on the scale lines so that the data fill up as much of the graph area as possible.
- Put tick marks outside the data region and keep them to a minimum.
- Label each scale line (quantity and units).
- Place the figure title and legend below the graph.

Logarithmic Graphs

While the line graph shows absolute differences or increments of change of a quantity, the logarithmic graph emphasizes relative or percentage change. For this reason, the logarithmic graph is an ex-

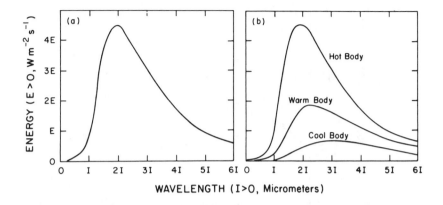

WAVELENGTH (I>O, Micrometers)

Figure 3. (a) Generalized shape of the "blackbody" emission curve for any body, showing the maximum possible emission from the body (at constant temperature) for each wavelength. The curve is described by an equation formulated by Max Planck in 1900 and is termed Planck's radiation curve. The equation is called Planck's law. (b) Comparative blackbody radiation curves for bodies of different temperatures. The warmer a body is, the greater is its blackbody emission at each wavelength and the shorter is the wavelength at which its emission peaks. (From W. M. Washington and C. L. Parkinson, 1986, *Three-Dimensional Climate Modeling,* Oxford, New York: Oxford University Press, p. 13. Reprinted with permission from University Science Books.)

cellent tool for comparing rates of change of different lines. A straight line that is ascending or descending indicates change at a constant rate, while a horizontal line indicates that the rate of change is zero. Parallel lines indicate the same rate of change. Thus, even though the absolute values of two series to be compared may differ widely or have different units, the slopes of the lines indicating rate of change can still be compared. See, for example, Figure 4, where the two sets of lines for the liquid and vapor phase of a methane-water system show first opposite rates of change and gradually approach equilibrium conditions, as indicated by the nearly horizontal lines at higher pressure.

Logarithmic graphs not only indicate relative changes, but show absolute values at the same time. They are, therefore, especially useful when the vertical scale goes through several powers of ten. However, in the logarithmic graph the absolute values cannot be visualized as distance from the baseline, but must be read off along the logarithmic scale. Since nontechnical readers often are not able to

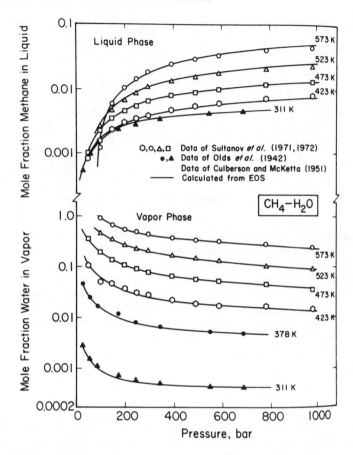

Figure 4. Vapor-liquid equilibria for the system methane-water. Liquid-phase isotherm at 378 K not shown for clarity. Binary parameters listed in Table. (From R. L. Cotterman and J. M. Prausnitz, 1986, "Molecular Thermodynamics for Fluids at Low and High Densities," *Am. Inst. Chem. Eng. J.* 32[11]:1804 [November]. Reproduced by permission of the American Institute of Chemical Engineers.)

interpret the logarithmic scale, these useful graphs belong in documents addressing experts or professionals.

Bar Graphs

The common bar graph is particularly appropriate for showing discrete values and setting them up for comparison. Bar graphs show well relationships between distinct quantities or the parts of a large group of data. They provide a clear display of the magnitudes of the various categories or parts and of their differences, and they allow the reader to scan the data rapidly and effectively. Figure 5, for example, shows a clear relationship between the size of remaining mobile oil reserves in Texas and the type of reservoir formation. Since bar graphs emphasize individual amounts rather than trends and directions, they have most impact when used to display relatively few values of one or more series. The most commonly used bar graphs are the simple bar graph, the clustered bar graph, the subdivided bar graph, and the subdivided 100-percent bar graph.

Unrecovered Mobile Oil in Texas

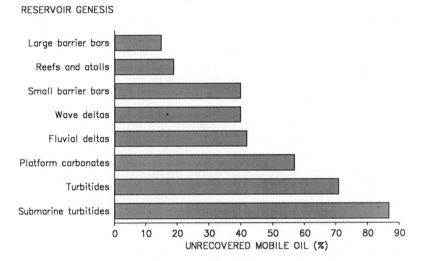

Figure 5. Unrecovered mobile oil as a function of reservoir genesis and associated macroscopic heterogeneity. Based on sample of 450 largest reservoirs in Texas. (Data from W. L. Fisher, 1987, "Can the U.S. Oil and Gas Resource Base Support Sustained Production?" *Science* 236:1634 [June].)

GUIDELINES for constructing bar graphs:

- Arrange the bars in a logical sequence.
- Use the same width for each bar, but make the distance between bars different from the width of the bars.
- Make the bars stand out from the white background.
- Since each bar represents magnitude by its length, include the zero line for accurate representation.

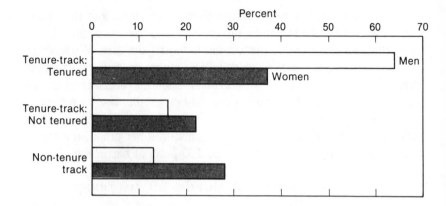

Figure 6. Doctoral scientists and engineers in educational institutions by tenure status and sex: 1981. Note: Detail does not add to 100 because "no report" is not included. About 13 percent of the women and 7 percent of the men did not report tenure status. (From *Women and Minorities in Science and Engineering*, 1984, Washington, DC: National Science Foundation, p. 6.)

- To gain vertical distance for the shorter bars, break an occasional, excessively long bar above the range of the other bars.
- Use vertical bars for comparing the magnitudes of a variable over time. Use horizontal bars for comparing the magnitudes of categories with descriptive labels.
- Label each scale line.
- Place the figure title and legend below the graph.

Clustered bar graphs. If you want your readers to compare the magnitude of two or more series of variables, you can display the data effectively by grouping bars into clusters. The length of each bar can readily be compared with the lengths of other bars in each group and the magnitude of each bar can also be estimated from the common standard scale. Figure 6, for example, highlights the discrepancy in hiring patterns for men and women in scientific and engineering fields. Each type of bar is distinguished by a different texture, and the clusters are spaced well apart. The bars can be labeled either in one cluster or in a key.

Subdivided bar graphs. In the subdivided bar graph each bar is further divided into components. The purpose of this graph is to allow a quick comparison of the relative sizes of a few components, in particular their relationship to the whole. For ease of comparison, well-constructed subdivided bar graphs as in Figure 7 show the largest or most important component at the bottom of the bar. Only the magnitude of this base component can be read off directly along the vertical scale. If the bars have several components above the base component, and if the magnitudes of these upper components are important, then the data should be displayed in separate graphs or in a clustered bar graph, with each component having its own bar.

Subdivided 100-percent bar graphs. In this graph each bar has the same length and represents 100 percent. Its purpose is to compare the individual parts of a whole and their relationship to the whole or 100 percent. The components of each bar, indicating percentage, are usually ordered according to size. However, when the various components are to be compared to some other common standard than the 100 percent, then individual bars for each component, grouped into clustered bars, will define contrasts much more clearly.

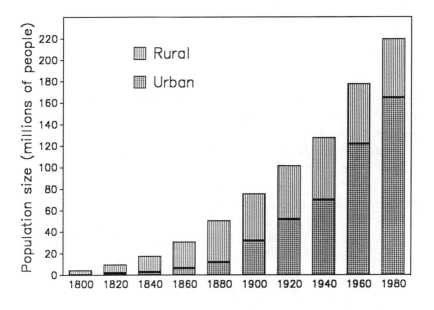

Figure 7. Growth of rural and urban population in the United States. (Data from P. R. Ehrlich and A. H. Ehrlich, 1970, *Population, Resources, Environment,* San Francisco: W. H. Freeman and Company.)

Where the money goes
(in billions of 1985 dollars)

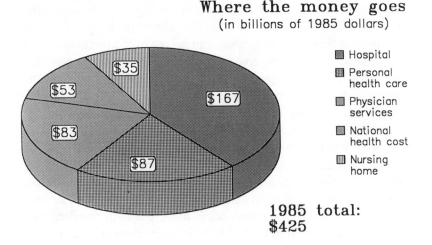

- ▦ Hospital
- ▦ Personal health care
- ▨ Physician services
- ☐ National health cost
- ▥ Nursing home

1985 total: $425

Figure 8. Who pays the nation's medical bills (in billions of 1985 dollars). (Data from "The Rising Costs of Health," *New York Times,* Sunday, 15 Feb. 1987, sec. 4.)

Pie Graphs

Pie graphs are 100-percent graphs and are widely used for showing the percentage distribution of the components of some body of data. Pie charts are especially useful for comparing percentages of money. Figure 8, for example, shows where the money goes for the nation's medical bills. Since it is difficult to accurately estimate angles or areas, pie graphs are intended to provide an overview rather than exact values. For this reason, pie graphs are widespread in literature for diverse audiences. You can make an important segment stand out by using special shading or by separating the segment from the main pie. You can also combine pie charts with line graphs to show percentages and actual data.

GUIDELINES for constructing pie graphs:

- Limit number of segments to five or six.
- Order from largest to smallest segment, beginning at twelve o'clock and moving to the right.
- Identify each sector with a label.
- Keep labels in segments, if possible, and keep labels horizontal.

Pictographs

Pictographs dramatize the facts and are designed to attract attention. They function like bar graphs in which the bars are replaced by a series of symbols that represent the quantity graphed. Each symbol represents a defined unit value. Since carefully chosen and easily recognized symbols add interest to statistical data, pictographs are successful with a general audience.

It is best to use only whole symbols rather than a fraction of, let's say, a tractor or a person, even though this practice results in generous approximations. After all, pictographs are meant to provide only an overall picture rather than minute details.

Flow Charts

While line graphs can guide the reader effectively through trends or changes in magnitude of one or more variables with time or changes in other variables, a flow chart can describe the steps or phases of a technical process or operation in graphical form. The flow chart highlights the sequence of steps or phases of the process and can specify the time required for these. Pictorial forms can add interest to these charts. A well-designed flow chart, as shown in Figure 9, can present a large number of facts clearly and simply, without extensive verbal descriptions.

Drawings

Drawings in a technical document serve numerous functions: they can help define an object by giving an overview, as in a technical description, and can clarify the relationship of the parts. Enlarged views, cutaways, or exploded views represent subdivisions of the whole into parts. In addition, drawings can illustrate or give examples.

Even though the actual drawing of illustrations is a technical skill and not part of the writing skill itself, the writer should consider how drawings serve certain writing strategies as well as make certain that the visuals meet effective design standards. The following guidelines for effective figure legends and for clear and effective visuals also apply to drawings.

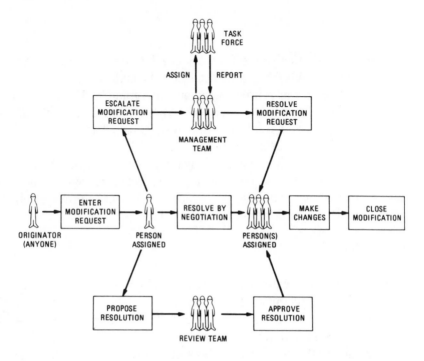

Figure 9. Resolution process for modification requests. (From T. S. Kennedy, D. A. Pezzutti, and T. L. Wang, 1985, "System 75, Project Development Environment," *AT&T Technical Journal* 64[1]:282 [January]. Reprinted with permission from the *AT&T Technical Journal.* Copyright 1985 AT&T.)

Writing Effective Figure Legends

Graphical displays not only supplement the text, but we have shown, often serve as invitation to reading the text. Because it appears to take less effort to decode graphical displays than to understand verbal sequences, readers often try to get by with reading only these parts of the text or at least to gain an initial understanding of the content and purpose. For example, when reading a proposal, the reviewer might concentrate on the graph that shows the schedule rather than on the details of the procedures in the text. For this reason, you do well not only in crafting your major conclusions in graphical form—be it a bullet list of recommendation, a flow chart of a modified production process, or a graphical illustration of engine performance—but also in complementing your visuals with comprehensive and informative legends. In fact, because these visuals are so often read independently from the text, you should strive to make the titles and legends complete enough so that the visual can stand by itself.

When composing a legend for a figure, you should first provide a clear idea of subject and purpose in the title. Then, in the legend, you should describe everything that is shown on the visual, point out what is important, and summarize what you conclude. The legend for Figure 3 is a good example. While the legend provides an explanation of the information in the visual, it should be brief and should not take the place of the text. Explanations of procedures, experimental set-ups, and details of data acquisition and calculation do not belong in the legend. But remember that engineers tend to provide too little detail in their discussions of visuals rather than too much. Consider, for example, Figure 10: the skimpy legend does not clarify the graph.

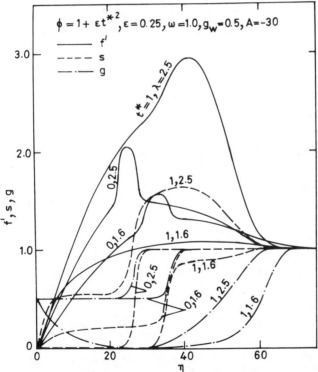

Figure 10. Effect of the Mach number/yaw/angle parameter on the velocity and total enthalphy profiles. (© Copyright American Institute of Aeronautics and Astronautics; reprinted with permission from R. Vasantha and G. Nath, 1986, "Boundary-Layer Flow Past a Cylinder with Massive Blowing," *Am. Inst. Aero. Astro. J.* 24[11]:1875.)

GUIDELINES for writing figure legends:

- Focus the title of the visual on its subject and purpose.
- Describe everything on the visual.
- Point out what is important.
- Summarize what you conclude from the visual.

DESIGNING VISUALS

In designing effective visuals we apply the same principles as in technical writing: we analyze our purpose and our audience, and we make important things stand out. Focusing on purpose and audience helps in selecting the type of visual that emphasizes the important point, and is appropriate and appealing to the reader. Making important things stand out helps to make the visual clear and accessible. Let's discuss how you can check your visuals for clarity and correctness.

Making Visuals Clear

One way to achieve clarity in visuals is to exclude anything that distracts and is not essential to the message of the graph. Thus, a major goal in drafting and revising a visual is to filter out nonessential information and nonessential graphical elements and embellishments. Let us consider these points more closely.

How much information should a visual include so that it is clear? The answer to this question derives in part from considerations of purpose and graphical perception and in part from audience analysis. First, visuals must be simple in that they present only one key thought. Confusion often results from trying to show too much. Since the viewer perceives a graph in its entirety rather than as a sequence, the information has to be organized so that the core idea stands out and is decoded first and without confusion. You may have much to say, but if you try to say it all at once, the message might not be understood. For example, to represent competing ideas in one visual, as in Figure 10, is bound to lead to confusion. The many lines do not readily reveal a dominant relationship. Of course, as seen in Figure 4, a visual can be clear even though it contains many data points. In Figure 4 all the lines have similar directions, indicating a clear relationship.

Second, everyone has experienced that what is clear to one person is not necessarily clear to another. Careful audience analysis will help to limit the complexity of the visual so that it is appropriate for the audience. For example, Figures 11–13 show diagrams of the greenhouse effect, perhaps the most commonly understood diagram illustrating an environmental principle. Only the first figure, though, is easy to understand for a general audience, and is indeed designed for a lay audience. The second figure, from the *Scientific American*, is designed for nonexpert, college-educated readers, and the third figure is a typical diagram of an undergraduate text in meteorology.

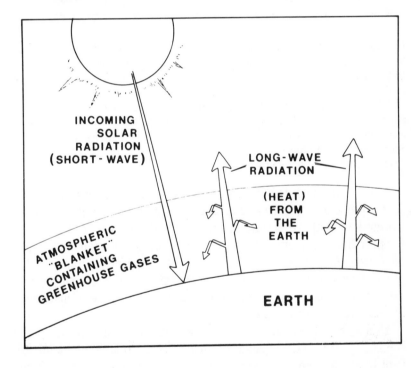

Figure 11. The greenhouse effect. Short-wave radiation from the Sun reaches the Earth's surface unhindered, but the outgoing long-wave radiation is partially trapped, or retained, by carbon dioxide and other gases in the atmosphere. (Reprinted by permission from F. P. Bretherton, 1986/87, "The Oceans, Climate, and Technology," *Oceanus* 29[4]:4.)

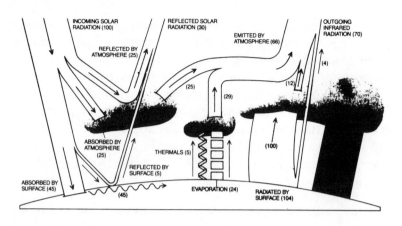

Figure 12. Greenhouse effect arises because the earth's atmosphere tends to trap heat near the surface. Carbon dioxide, water vapor and other gases are relatively transparent to the visible and near-infrared wavelengths (med. gray) that carry most of the energy of sunlight, but they absorb more efficiently the longer, infrared wavelengths (lt. gray) emitted by the earth. Most of this energy is radiated back downward (dark gray). Hence an increase in the atmospheric concentration of greenhouse gases tends to warm the surface. (Reprinted by permission from S. S. Schneider, "Climate Modeling," *Scientific American* 256[5]:78. Copyright 1987 by Scientific American, Inc. All rights reserved.)

Just as too much information makes graphical displays less clear, so do too many graphical elements and embellishments. In other words, just as there is cluttered writing, there is cluttered graphics. Chapter 3 gives many examples of passages overloaded with show-off words, filler words, or empty phrases. Similarly, graphical displays of technical material are often cluttered with distracting hatching, overabundant tick marks, or too many labels and explanations in the data region of the graph.

To avoid graphical clutter and to keep graphical displays clear and simple, E. Tufte (1983) suggests that non-data-related elements are kept to a minimum. Accordingly, you will create better visuals if you use thin grid lines or no grid lines at all, and if you help the reader to focus on the information by avoiding distracting optical vibrations produced by carelessly chosen diagonal hatching. You can improve your visuals by simplifying tick marks and labels, by reducing the graph area to the size of the data region, and by placing explanations and lengthy keys outside the graph.

This advice for avoiding graphical clutter is helpful for drawings as well. Consider, for example, the use of cartoon techniques in the

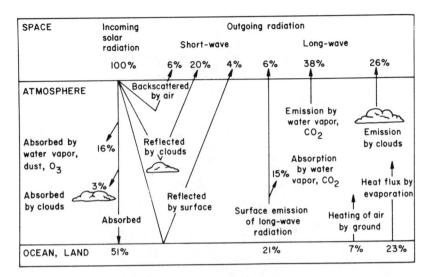

Figure 13. The radiation budget of the Earth. Most of the energy from the Sun arrives in form of short-wave (visible) radiation. Some of it is reflected into space by air molecules, clouds, and the Earth's surface. The rest is absorbed and heats the Earth. The warm Earth emits long-wave (infrared) light, part of which is absorbed in the atmosphere. The retention of infrared radiation by the atmosphere leads to the "greenhouse effect." (Reprinted by permission from M. O. Andreae, 1986/87, "The Oceans as a Source of Biogenic Gases," *Oceanus* 29[4]:32.)

safety warning in Figure 14 that makes understanding possible only on second thought, and the simple, but vivid and dynamic lines of Figure 15 that clearly focus attention on the important message of keeping hands out.

Making Visuals Correct

Visuals are sometimes designed to conceal the truth. For example, an aggressively designed advertisement might exploit certain fallacies in visual perception. However, in technical documents such misuse of graphics is not permissible and usually not intentional. In the long run, to remain credible you must come across as objective and exact in your communication. The following discussion should help you to avoid unintentionally fuzzy and misleading visuals.

Research of graphical perception has shown that we perform some tasks of decoding graphical forms more accurately than others. W. S. Cleveland (1985) reports that viewers judge best length against a

 CAUTION! Keep all shields in place at all times.

Figure 14. Warning symbol. (Reprinted by permission from *Operator's Manual, International 3160 Series A Rotary Mower,* Setting Up Instructions.)

common scale and that they can judge length better than angles or slopes and, least of all, volume and color. The list below orders the common graphical perception tasks from most to least accurate.

1. Position along a common scale.
2. Position along identical, aligned scale.
3. Length.
4. Angle-Slope.
5. Area.
6. Volume.
7. Color hue—Color saturation—Density.

Cleveland advises that in encoding data we should ensure that the viewer will do the decoding of important information using judgments

Figure 15. Warning symbol. (Reprinted by permission from FMC Corporation, 1985, *Product Safety and Label System.*)

based on tasks in the upper part of the list. For example, a line graph or simple bar graph is more reliable than a pie graph because the important information is judged by position along a common scale rather than by angle. More subtle is the case of the subdivided bar graph in Figure 16. Even though the graph seems compact and attractive, it is harder to understand than in a line graph or a simple bar graph because, instead of position along a common scale, the viewer must decode length for all but the base component. Although the reader can judge accurately only the component at the base of each bar, he or she is asked in the legend to focus on the upper two components. Apparently for this reason, the author added the dollar amount in each component; however, the many numbers make the graph look cluttered. A clustered bar graph would have been the better choice.

Angle judgments are even less accurate than length judgments. They are necessary for reading pie graphs, which are so frequently

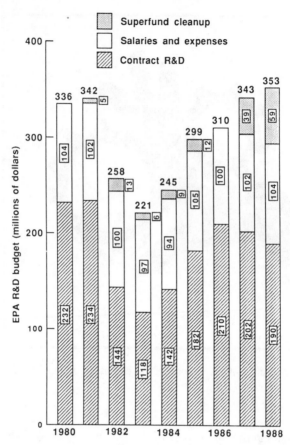

Figure 16. R&D budget. Funding of research to support Superfund cleanup sites would increase $20 million in fiscal year 1988, while "salary and expense" money for in-house research and contract R&D funds decline. (Reprinted by permission from M. Crawford, 1987, "R&D Eroding at EPA," *Science* 236:905 [22 May]. Copyright 1987 by the AAAS.)

used in business and by the mass media. Readers, in general, feel comfortable with pie graphs, perhaps because many have learned to visualize percentages and fractions by dividing real apple pies among the family members. However, let us also remember the fights and arguments about who got the biggest piece. For good reason, therefore, engineers shy away from using pie graphs in technical reports.

Still less accurate than angle judgments are area and volume judgments. Even though graphs using area or volume for decoding data, such as pictographs, add interest and draw attention, they should be reserved for general audiences and should be decoded with caution.

Another important consideration in graphing data accurately is that of choosing appropriate scale lines. The range of tick marks along the scale lines should be chosen so that the graph area defined by the scale lines is that of the data region. Limiting the graph area in this way may sometimes mean that the graph does not include the zero point on either of the scale lines. For example, the plot of the Dow Jones average for one month in Figure 17 as commonly published in daily newspapers does not include the zero point for the vertical scale, but only starts at 2250. In a way, the graph resembles the outline of the top of an iceberg—the part that's above the water. Yet, most viewers are interested only in this outline; you can assume that they are familiar with the absolute magnitudes and will not misinterpret the data. The same graph, but with the scale line starting at zero, consists mostly of wasted space and shows reduced resolution of the actual data points that carry the most important information for the viewer. Suppressing the zero line in cases like this is preferable because emphasis in line graphs is not on absolute values but precisely on trends. Similarly, in bar graphs the zero can sometimes be suppressed to avoid wasted graph area.

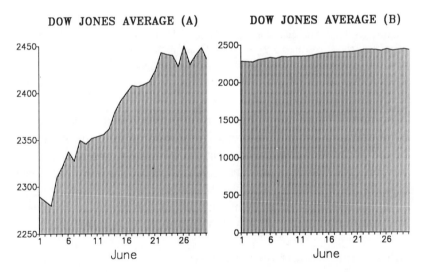

Figure 17. Dow Jones Average for June 1987. (A) The graph area is limited to the data region. (B) The graph area includes the zero point of the vertical scale line. (Data from *Wisconsin State Journal*, 7 July 1987, sec. 3.)

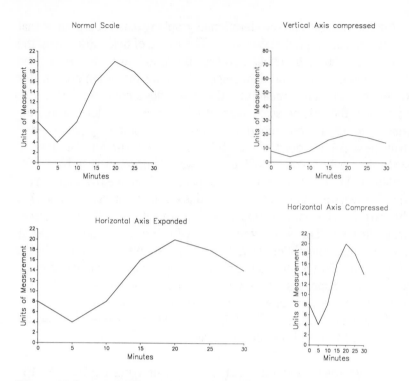

Figure 18. Impact of scale unit length on slope. Manipulating the units of measurement along the x-axis makes the slope of the line or data trends appear more or less dramatic. Manipulating the units of measurement along the y-axis makes the range of data appear more or less prominent.

In addition to the range of the scale, the choice of scale unit lengths is important because it influences the slope of lines. As shown in Figure 18, condensing the horizontal scale (x-axis) or expanding the vertical scale (y-axis) will accentuate slopes. Therefore, if small changes are important, you should choose the scales so that these changes will stand out as significant, just as you would use an explosive view or an enlarged view in a drawing if detail is important.

Integrating Visuals into the Text

To help your readers get the most out of your visuals, you should strive to integrate graphical displays verbally into the text itself. The visual will better supplement the text if you refer to it by figure number as soon as you begin discussing it rather than after you are done with the discussion. Furthermore, even though the legend already explains the figure, you should provide a complete explanation in the text also. You would inconvenience your reader if you send him back and forth between figure legend and text. Finally, you should place the figure as near as possible to its discussion, preferably on the same page. Only the highly motivated reader will search for figures in the appendix.

REFERENCES

Cleveland, W. S. 1985. *The Elements of Graphing Data.* Monterey, CA: Wadsworth Advanced Books and Software.
Tufte, E. R. 1983. *The Visual Display of Quantitative Information.* Cheshire, CT: Graphic Press.

Chapter 5

The Computer Revolution in Writing

Engineers and scientists once used drafting boards, calculators, and typewriters in their day-to-day office work. Now all of you use computers instead of calculators for complex calculations; some of you use computers in place of drafting boards; and increasingly large numbers of engineers and scientists use computers instead of typewriters to produce written documents. So great has the impact of personal computers on writing been that one hears frequently of "the computer revolution in writing."

The book you are reading is largely a product of that computer revolution in writing. We would never have had the time to undertake the project had we not had personal computers and sophisticated software to help us plan, organize, store, format, edit, and print the information that we wanted to present. Computer technology helped us to write and revise outlines and drafts without the organizational load imposed by trying to get from rough drafts through a typist to final drafts ready for publication. Instead, we put information directly into a computer's memory where we could manipulate it and retrieve it much more conveniently.

Given all that computers have done to make our own writing tasks easier, we were tempted to make this chapter one long public service announcement designed to convince all engineers and scientists who don't already do so to write on a computer. Certainly, computers are an appealing choice for engineers and scientists looking for a way to get more out of the time they spend writing. The engineers and scientists we know have access to computers, have at least rudimentary keyboard skills, and have experience in using computers to store and process information.

Yet, we resisted that temptation. We feared that in the process we might mislead some to think that using a computer for writing is a magic bullet that slays all writing problems. It isn't, any more than a computer is magic in any other sense. All computers can do is help streamline the writing *process*. Their effect on the *product* is not direct or obvious. A decade of research on the effect of computers on the composition process has disappointed those—including many English teachers—who looked at the use of computers as an easy way to improve writing skills.

Instead, then, we will try to describe the sometimes overwhelming array of computer software and hardware that is out there that might make your writing process *more efficient* and the final product *more attractive* and *more easily revised*. The chapter ends with a bibliography on the subjects of computer-aided writing and computer-aided (or "desktop") publishing. When armed with the knowledge in this chapter, you are encouraged to keep current with the field by occasionally reading one of the many weekly and monthly publications on personal computers. We also encourage you to contact a local computer users' group, whether you own a computer or are interested in buying one. You will meet people who can give you practical answers to specific questions you might have on either starting or expanding your use of computers for writing.

WHY ARE THE CHOICES SO OVERWHELMING?

The variety of computers, word processing–related software and printers attests to the great variety in the needs of users (and the great number of bright young computer professionals). We see these three categories of users:

> **Level One**. Some engineers and scientists have no special needs. They just need to create correspondence and short reports. Since these documents don't require much in the way of formatting, those who write them don't want to buy an expensive system or learn a program complicated by extra features that will never be used.
>
> **Level Two**. Other engineers and scientists need to combine text and graphics in the same documents. These writers know that they can use the same scissors and paste that they have always used to merge writing and pictures. However, they also know that since the

writing and pictures are both more easily produced with the computer, it makes more sense to have the computer combine the two before printing.

Level Three. Finally, some engineers and scientists need to produce formal documents like monographs, annual reports and instruction manuals—the sorts of things that traditionally have been sent to professional typesetters for layout and production. For a variety of reasons (lower cost, shorter lead-time, increased control), these users want to be able to produce attractive, professional publications in their own offices.

The diversity in the needs of users extends much farther in so many directions that further categorization is impossible. Some engineers and scientists need to be able to send and receive information to and from other computers. Some want to be able to switch quickly between spreadsheet or database programs and their word processing program as they calculate and collect information for whatever they are writing. Others would like to use the computer's information processing ability to help correct spelling, punctuation, and stylistic errors in writing. Some users write almost as many equations as they do sentences. They need help with the typesetter's nightmare of subscripts, superscripts, integrals, Greek characters, and special symbols that the ordinary word processing program just isn't able to handle. This diversity of needs partially explains the overwhelming number of commercially available hardware and software products, but it doesn't make it easy for engineers or scientists to find out what to buy so that they can do what they want to do. We hope the following sections give you ideas for what might suit your particular writing and publishing needs.

The first section gives advice on *how to buy*. The next sections discuss the computer *hardware, software, and printers* that correspond to the different user needs. These sections turn out to be organized chronologically as well; the section on word processing basics describes the oldest (circa 1980) technology, while the last describes the newest. Finally, we describe the *miscellaneous products* that are meant to make your writing easier or better in one way or another.

Throughout, we mention brand names because the information wouldn't be as helpful without them. However, since no one has paid us to plug a particular product, we can claim total neutrality: any bias you might perceive is ignorance or oversight in disguise.

HOW AND WHAT TO BUY

Computing power is so cheap relative to what was available even five years ago that it seems almost unnecessary to worry about how or what to buy in the way of computers. Nonetheless, some care should be used to make sure that you will be happy with whatever you buy to expedite your writing. The key words to keep in mind are *compatibility* and *expansibility*.

Compatibility is important because you will want to have as large a selection of software and add-ons (printers, memory expansion units, communication devices) to choose from as possible. Today that means that you will probably want to consider an IBM personal computer; a computer that is compatible with the IBM, such as a Compaq or Zenith; or an Apple personal computer, most likely an Apple Macintosh. If you opt for an IBM compatible, you will pay less than you would for the real thing; however, be aware that compatibility is a matter of degree, and that you are likely to find software (especially graphics software) that will work well on an IBM that will not work at all on your compatible. The trick is to find an IBM-compatible computer that will run those programs that you foresee using.

Expansibility is important because you will undoubtedly find new uses for your computer as your sophistication increases. Make sure that your computer is designed so that it can be upgraded, especially in terms of memory and input devices. Two of your authors own expansible computers, and are very happy with them. One doesn't, and isn't.

Once you have decided what you need, you have a clear-cut choice to make: whether to buy locally or from one of the mail-order houses that sells whatever you can buy from your computer dealer at a lower price. Your choice depends, of course, on how much you value a dealer's service and help; a mail-order house won't be of much help if you have trouble setting up a new system. Your choice also depends on how much you trust advertising. As with other mail-order purchases, fraud is a possibility if you respond to an ad that sounds to good to be true. Since most major computer magazines (such as *PC Week* and *MACazine*) only accept advertising from reputable merchants, you should feel confident that you won't be blatantly swindled if you buy from an ad in a major magazine. On the other hand, you may not get exactly what you expect, and you will have little recourse if you don't. Once again, contacts at a local users' group might help you decide from whom to buy.

Level One: Word Processing Basics

If your needs are modest and your budget small, you are looking for basic, easy-to-use hardware and software. You probably would like to buy cheap and get started fast. Since you are not after the latest in technology, you shouldn't have any trouble finding what you need.

Hardware

Any personal computer with at least 128K, one or more floppy disk drives, and the ability to send 80 columns of text in one line will do the job. Amiga, Apple, Atari, Commodore, and Tandy are brand names of personal computers that sell for less than $1,000. They differ in many ways, but for the purposes of a word processing engineer or scientist, criteria such as which computers your friends own or how the keyboard feels to you should lead you to a purchase you can live with.

The monitor ("television" or "cathode-ray tube" display) is often sold separately. Spend more time choosing it than you did the computer. Basic word processing doesn't require color, but it does require a monitor that can show 80 columns of text clearly. An ordinary television set can show the output from a computer, but it will make you bleary-eyed if you try to look at the grainy, flickering display for the time needed to write even a short document. On the other hand, small monochrome (usually green or amber) monitors with crisp, contrasty displays sell for less than $150. Any one should work with almost any small computer.

Software

With only a few exceptions, you will have to purchase word processing software separately. (Exceptions are special promotions or odd product offerings that include a word processing program with a computer.) Because for the most part each personal computer has its own operating system (that is, its own procedures for handling information), you will have to shop for a word processing package that is labeled as being for your computer.

What should you look for in a low-cost, easy-to-learn word processing program? Five years ago that was a burning question, because products differed greatly in usability and power. In the meantime, intense market pressure has eliminated the poor programs and forced the mediocre ones to improve. Today, for less than $75 you should be able to buy a word processing program (such as My Word!, for IBM and compatibles—$49 without spelling checker, $59 with spelling checker) that will allow you to do all of the following:

- Create and store documents of up to 20 pages.
- Customize the appearance of what will be printed.
- Find or replace any particular word or string of characters (such as the current date) wherever it might appear in the document.
- Move through the document with electronic speed so that you can make editing changes as needed.
- Create some special printing effects, like underlined or bold-faced characters.
- Merge information from one document into another.
- Move or copy blocks of text—that is, do electronic "scissors-and-paste" editing.
- Send a document to one of several common printer models for printing.

By today's standards, these are prosaic feats, and all products on the market should do them with ease. While they are now commonplace features, they are nonetheless impressive talents when compared to the abilities of the document-processing standard of ten years ago, the IBM Selectric typewriter.

While they all let you do about the same tricks, without question low-cost word processing programs vary greatly in ease of learning. Some, called "command driven," require you to learn quite a few keyboard commands, while others, called "menu driven," display screens of choices from which you simply choose the appropriate response. Menu-driven programs, of which Bank Street Writer is a commonly available example, are of course much easier to learn. The advantage of the command-driven programs becomes apparent weeks or months later when you know exactly what to do, but still have to wade through layers of menus to get the job done. The solution to this dilemma of menus versus commands is either to buy a program that can be controlled either way, or to buy a program that has extensive on-line help in the form of screens of information that you can read when you get stuck trying to figure out which keys to hit.

Printers

All word-processing products have come down in price even as their quality has been improving, but those trends are most evident in the field of low-cost printers. Therefore, buying a printer that you can live with has become quite easy.

For $200 to $350, you can purchase a reasonably versatile dot-matrix printer that will out-perform printers that cost over $1000 just five years ago. All of these prodigies, which are made by manufacturers such as Epson, Panasonic, Okidata, and Star, print faster, more quietly, and more reliably than their lumbering ancestors, which mimicked miniature washing machines both in appearance and decibel level. The best-selling inexpensive printers all create characters on regular paper using nine closely spaced pins that contact a typewriterlike ribbon. Figure 1 shows a sample of the print produced by several inexpensive printers.

Choosing among these brands is a matter of weighing minor differences in price, print quality, speed, service availability, and quality of documentation. Of these criteria, speed might seem the most discriminating, but unfortunately the manufacturers in this competitive field have taken to publishing speed ratings that can be believed only by those who still take EPA gas mileage figures seriously. Other criteria that should influence a buyer's decision include graphics capabilities and carriage width, since some users may want a wider carriage to print spreadsheets.

Since all of the printers in this price range are bargains, a buyer should concentrate instead on determining how well the printer will work with the already chosen computer and word processing software. In particular, a buyer should study the word processing program documentation to see which printers the program can communicate with. Buy one of those printers. Other printers would probably work, but compatibility of printers and software is definitely a matter of degree. Nothing is more frustrating than having a program that can do some fancy trick that the printer could do if only you could figure out how to get the two to communicate. Avoid that frustration by choosing a program and a printer that know how to work together.

We wrote this short book on writing so that you, as an engineer or scientist, could have the insights and strategies necessary to be analytic about writing. After reading it, you should, with your new classical understanding, be able to write better in every respect: more quickly, more clearly, more convincingly. Our major premise is that anyone who is careful and analytical can write well when provided with some writing strategies and insights.

We wrote this short book on writing so that you, as an engineer or scientist, could have the insights and strategies necessary to be analytic about writing. After reading it, you should, with your new classical understanding, be able to write better in every respect: more quickly, more clearly, more convincingly. Our major premise is that anyone who is careful and analytical can write well when provided with some writing strategies and insights.

We wrote this short book on writing so that you, as an engineer or scientist, could have the insights and strategies necessary to be analytic about writing. After reading it, you should, with your new classical understanding, be able to write better in every respect: more quickly, more clearly, more convincingly. Our major premise is that anyone who is careful and analytical can write well when provided with some writing strategies and insights.

We wrote this short book on writing so that you, as an engineer or scientist, could have the insights and strategies necessary to be analytic about writing. After reading it, you should, with your new classical understanding, be able to write better in every respect: more quickly, more clearly, more convincingly. Our major premise is that anyone who is careful and analytical can write well when provided with some writing strategies and insights.

Figure 1. Samples of output from standard dot-matrix printers. From top to bottom: the IBM Proprinter (draft mode); the IBM Proprinter (NLQ mode); the Epson FX-80 (draft mode); and the Dataproducts 8050 (draft mode).

Interface

A fresh-out-of-the-box computer cannot communicate with a fresh-out-of-the-box printer because two components are missing: the interface and the cable that links the two together. Unless you are a computer hobbyist or electrical engineer, neither is interesting. However, both are necessary.

The interface resides inside the computer cabinet, yet quite frequently it must be purchased separately. It is a circuit board that mounts near the computer's back panel; one of its components is a socket to which one end of the cable attaches.

Interfaces are of two distinct types, *parallel* and *serial*, each of which has distinct advantages. Parallel interfaces send information rapidly over eight or nine wires that connect to the printer. Serial interfaces send information less rapidly and somewhat less reliably over a single wire or pair of wires. The advantage of serial interfaces is that serial transmission can take place over thousands of miles, while parallel transmission can only occur over a few feet. Since we are talking about connecting a computer and printer that sit on the same desk, the parallel interface is obviously the better choice. A better choice yet would be a parallel interface built into the computer, but typically you will have to spend $80 to $200 for the privilege of installing one yourself.

Now all that is left to discuss, buy, or figure out is the cable that leads from the interface to the back of the printer. Typically, neither your computer nor your printer will come with one. While you could probably wade through the printer documentation and figure out how to engineer a cable that would work, the cost-effective alternative is to go to a computer store and tell them that you want six feet of cable to connect [your computer brand] with [your printer brand]. You will leave about $20 poorer than you came, but you will have saved time and avoided anguish.

Level Two: Combining Text and Graphics

Whether the writing project be an article for publication or a final engineering report on a consulting project, text and graphics must combine for satisfactory results. Engineers and scientists who needed to produce both in a single document have radically changed the development of microcomputers.

As an historical aside, word processing and computer-aided graphics evolved simultaneously, but quite independently. As of about 1976, programs became available that allowed innovative graduate students (usually in computer science) to use the number-crunching behemoths such as the IBM 370 to help them write their theses—sometimes punched line-by-line on IBM cards. At about the same time, software evolved that drove large, expensive Houston plotters; with sufficient programming knowledge and a day's time to fool around with the balky pens, a graduate student could engineer a bar or line graph that was suitable for publication.

Until about 1982, even enterprising graduate students had not figured out software that could produce both text and graphics simultaneously. The difficulty can be explained by the radical difference between how computers have traditionally handled graphics and text characters.

Put quite simply, a computer must produce, store, and recreate any graphic bit by bit—that is, every point on a drawing on the computer screen must have a corresponding bit in the computer's memory that is either on or off. The first generation of microcomputers, as typified by the Apple II, was up to the task, but the limited memory, slow processing times, and grainy screen resolution made graphics a marginal proposition.

Handling text has always been much easier. Built in to the computer's nervous system is a routine for converting each keyboard entry into a code that allows the computer to handle each character as a unit—that is, byte by byte. The character **a** is represented by a certain byte that results in the correct matrix of dots being printed on the screen. So different are the bit-by-bit and the byte-by-byte approaches that microcomputer memory is usually divided into two different segments: graphics and text. It is very much a never-the-twain-shall-meet situation.

How the Macintosh Has Changed Word Processing

That situation changed with the 1983 introduction of the Apple Macintosh, an enormously successful offspring of the enormously unsuccessful Lisa. Designed around a much faster, more powerful central processing unit (CPU), the Macintosh brought text and graphics together by handling both bit-by-bit, a technology that has come to be known as *bit mapping*. While previous CPUs, which handled only eight bits at a time, could not store and revise screen displays quickly enough to keep up with an active user, the Macintosh's 16-bit CPU allowed continuous bit-mapping of text and graphics input.

The Macintosh has changed the look and feel of microcomputing. Users control its operation by responding to icons instead of printed menus; by pointing to graphics with a hand-held mouse instead of entering command characters or pressing cursor keys; and by using software that allows text and graphics to be intermixed freely. Because characters are not printed on the screen or on paper according to a predetermined code, they can be shown in any one of many different fonts (see Figure 2). The Macintosh approach to producing text and graphics, which has so greatly simplified their production, has resulted in, among other things, an overabundance of eye-catching posters on college campuses and homemade greeting cards with computer-generated Old English text and reindeer. More important to engineers and scientists are the gains in productivity and the impressive-looking results that the Macintosh yields without the usual computer-related bogies of long learning times and significant capital investment.

The Macintosh, then, in any of its present models (Macintosh Plus, Macintosh SE, or Macintosh II), allows an engineer or scientist to produce *sans* scissors and glue those final engineering reports and articles that contain both text and graphics. The basic Macintosh with MacWrite, MacDraw, and MacPaint software, and the Apple ImageWriter printer, which together might cost $2500, can do what the computing giants of the last decade couldn't do: handle graphics and text in a coordinated fashion. That feat shouldn't be too surprising, since the Macintosh has a faster CPU and more internal memory than some mainframe computers of 10 to 15 years ago.

How the Competition Has Responded

So far it may sound as if the Macintosh is the personal computing Goliath. The situation is really much more complicated than that. Up until about 1985, the IBM PC was cast in the Goliath role; Macintosh played the sling-wielding David; and all the other manufacturers played the bit parts of the minor warring factions. That situation has changed with the sales spurt of IBM PC XT and AT "clones," computers that mimic (and sometimes surpass in speed) the operation of

Chicago, 9 point: Now is the time for the man of the world to do that for which he is
Geneva, 10 point: Now is the time for the man of the world to do that for which he is
Monaco, 12 point: Now is the time for the man of the world
Bookman, 12 point: Now is the time for the man of the world to do
Zapf Chan, 12 point: Now is the time for the man of the world to do that for which
Times, 14 point: Now is the time for the man of the world to do that for
Helvetica, 14 point: Now is the time for the man of the world to do that

Courier, 10 point: Now is the time for the man of the world to do that

Σιμβολ, 18 ποιντ: νοω ισ τηε τιμε φορ τηε μαν οφ

New Cent... , 12 point: Now is the time for the man of the world to do that

Helvetica 24 point, **bold,** *italic,* underline,

outline, shadow, SMALL CAPS, superscripts,

and subscripts!

Figure 2.   Sample of fonts and type sizes available on the Macintosh, using Microsoft Word 1.05 software and an Applelaser Plus printer.

the IBM originals. Now IBM-compatible machines outsell IBM products. Together they outsell Macintosh products about four to one,
with Macintosh gaining slowly but steadily in its market share.

Rather than slaying the IBM Goliath, Macintosh has instead influenced greatly the direction of the evolution of the entire microcomputer field. Since it was introduced in 1982, the IBM Personal Computer, its documentation, and the software written for it have shown
IBM's mainframe-computer heritage. As an example, the Personal
Computer's power switch is marked with the binary characters **1** and
0 rather than **on** and **off**. More significantly, its disk operating system (DOS, the collection of commands by which a user controls basic
computer functions) is traditional in that it accepts certain alphanu-
meric characters as commands and spits back other characters in
response. Often these DOS commands require complicated syntax
and punctuation. Further, most word processing programs that have
become IBM standards resemble their mainframe ancestors in their
total disregard for users who wish to incorporate graphics.

The success of the graphics-oriented Macintosh has turned the
heads of those who develop software for the IBM and its clones. As a
result, engineers and scientists who wish to produce integrated text
and graphics (and, as a corollary, control programs with a mouse in
response to icons rather than with command characters in response

to prompts) on the IBM can now do so through one of several approaches, none of which is as tidy as the Macintosh solution.

One approach is to buy what has become known as an integrated software package. The IBM Assistant Series, which comprises the Writing Assistant (word processing), Graphing Assistant, Filing Assistant (data base manager), and Drawing Assistant, allows text and graphics to be intermingled and printed as a single document on an IBM Graphics Printer.

Another approach is to patch together text from a word processing program (such as Wordstar, WordPerfect, OfficeWriter, or Volkswriter) and computer graphics from any source (such as Lotus 1-2-3, the most common source; AutoCad, a microcomputer drafting program; or even BASIC-language-generated graphics) into one document by using a third piece of software called Inset. This program, which supports only a limited number of printers, allows graphics to be expanded, reduced, or cropped to size. It then works alongside the word processing program to control placement and printing of the graphics within a document. Figure 3 shows an example of text generated with Volkswriter Deluxe merged via Inset with AutoCad drawings.

A third approach is to use a program such as Microsoft Windows which gives IBM users a bit-mapped, mouse-controlled way to control switching back and forth between programs. In short, it makes IBMs much more Macintosh-like, but it requires extra internal memory to do so.

Thanks to the influence of the Macintosh, then, engineers and scientists who want to mix text and graphics using an IBM system can now do so. They will have to pay a bit more than those who opt for a Macintosh system, though, especially if their brand loyalty is strong. As of this writing, the IBM Personal Computer is no longer being made; thus, the cheapest IBM personal computer is the Personal System/2 Model 25. It and an IBM Proprinter, along with word processing, graphics, and Microsoft Windows software, cost over $3000. A package of IBM-imitation products would cost much less but still allow production of text and graphics. The Leading Edge computer (a popular foreign-made IBM clone that comes with its own word processing program), a Toshiba P341 printer, and graphics and Inset software might cost $2300.

In our' survey, we found that students prefer to choose their entree first and build their meal around it. They like to choose their salad second and then choose their drinks third. Typically, they choose dessert fourth and then pick up napkins and silverware last.

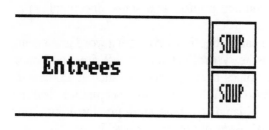

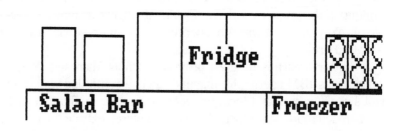

Fig. 1.   Proposed layout for Chadbourne dining hall

SOLUTION
     A variety of solution algorithms were considered including simulation and stochastic queuing.   The algorithm that we decided upon, however, was heuristic in nature but it contained elements of cluster analysis, flow sequencing, and human factors. We applied this information by making suggestions to

Figure 3.   Page of text interspersed with graphics. Produced with Inset, Volkswriter Deluxe Plus, and AutoCad software on a Dataproducts 8050 dot-matrix printer.

Printing Graphics and Text

While improved CPUs and cheaper memory have proven up to the task of providing integrated text and graphics, keeping printer technology abreast of the rapid changes has proven to be much more of a challenge for the microcomputer industry.

The traditional technology of dot-matrix impact printers has obvious mechanical drawbacks that limit both resolution and speed. Resolution can be improved by engineering finer print head wires, but the result is a more fragile product. Resolution can also be improved by a second pass of the print head to fill in the matrix more fully; speed obviously suffers as a result. The final method for improving resolution is to resort to the alternative technology of fully formed characters ("daisywheel" printing); this solution of course prevents printing of graphics of any kind.

More graphics in word processing files put an even heavier workload on dot-matrix printers. While the new CPUs can display a screen bit-by bit in a fraction of a second, to print that same screen requires as much as several minutes. This tortoise-like pace is, of course, out of step with every other aspect of today's word processing technology. As mentioned previously, low-cost computer printers have improved enormously in the last five years, but even the latest dot-matrix printers, those built with 24-pin print heads, are a bottleneck in the modern processing of words—and graphics.

The microcomputer industry has within the last two years developed a solution: print all the dots at once instead of just nine to twenty-four at a time. The technology for doing so already existed: the photocopying machine. In 1984, Hewlett-Packard, with the help of the copying expertise of Canon, produced the first laser-based page printer: the HP LaserJet. By applying the optics and printing technology of photocopying to the problem of printing text and graphics, Hewlett-Packard marketed a printer that could print a whole page of text and graphics in seconds instead of minutes, with a resolution of 300 dots per inch instead of 100, and with a gentle hum rather than the clatter of wires colliding with a platen at high speed.

Faster, clearer, and quieter: the LaserJet and its offspring have jarred our thinking about automated production of writing just a few years after the initial impact of the idea of word processing itself. The result is the mini-revolution of desktop publishing and the resulting third level in sophistication of word processing users: those who produce publication-quality text and graphics without the time and expense of traditional offset printing methods.

Level Three: Publication-Quality Text and Graphics

You have heard, we are sure, of desktop publishing, which we will refer to as DTP to avoid overusing what is a confusing term to start with. It is what people talk about at cocktail parties when the conversation strays toward high-tech subjects. It is the subject of frequent seminars, books, and even periodicals (see the bibliography at the end of this chapter). We have even heard that it is to the Information Age what Gutenberg's printing press was to the Renaissance.

Apparently, it is above all a topic that lends itself to exaggeration. But, lest we get ahead of ourselves, it is a term that, however common it might be, stands in need of definition.

According to one of the books on DTP (a book produced by the technology that is its subject), desktop publishing is "the use of personal computers to compose and print professional quality documents" (Ritvo and Kearsley, 1986). The desktop part of the term comes from two different angles:

• Unlike traditional printing presses, laser printers and the computers that drive them do fit on top of a table.
• Unlike traditional printing presses, desktops are everywhere; we all have one. Advertising types would like us to think that our desktops are not complete without printing machinery sitting on top of them.

In practice, DTP requires more than just a computer, a laser printer, and a word processing program. First, special software for laying out pages is required. It handles text and graphics from different sources, and gives control over print size and style. A typical example is PageMaker from Aldus Corporation. Second, some supercharging of the basic microcomputer is required. To process documents of journal or book length requires additional internal memory (usually totalling 640K to 1 megabyte) and a hard-disk drive to speed up information retrieval. Third, a high-resolution viewing screen such as the Wyse 700 is needed to take full advantage of the layout capabilities of the software. A final DTP component, which, although not essential, is certainly useful, is a scanner, a device that transforms any image into a digital map in the computer's memory. The scanner allows page layout software to put a photograph into a newsletter, for example.

DTP does more printing tricks than most engineers and scientists will require, unless they write user manuals or flashy proposals, or prepare camera-ready copy for journals. Since the technology is changing so fast that we can't describe DTP components and capabilities completely, we will describe three typical systems: a Macintosh-based system, an IBM-based system, and an off-the-shelf, completely packaged DTP system sold by Xerox. An interested reader can find more information on other DTP hardware and software by consulting the bibliography at the end of this chapter.

Macintosh-Based DTP System

With its visually oriented operating system, the Macintosh has been a natural starting point for development of DTP technology. To capitalize on the Macintosh's strength in DTP, Apple introduced in 1986 the Macintosh Plus and the LaserWriter, a versatile combination that should keep Apple ahead of the DTP pack. Figure 4 shows a page from a user manual that was produced on a Macintosh.

Table 1 lists components that you could purchase to set up a Macintosh-based DTP system. (Prices are approximate discounted retail.)

IBM-Based DTP System

IBM microcomputers, while standard setters for business applications, have until recently been ignored in the development of DTP technology. That trend has changed with the introduction of the IBM AT and its faster CPU, and the even faster 80386 CPU chip that is built into the top-of-the-line IBM Personal Computer System/2 Model 80. Adding to that momentum has been the announcement by IBM that it would adhere to the standards set by PostScript, the instruction language that is spoken by PageMaker (and other Macintosh page layout programs) to laser printers. A further factor that all along has worked in IBM's favor has been the many Lotus 1-2-3 users who would like to be able to print 1-2-3 graphics in their DTP-produced documents.

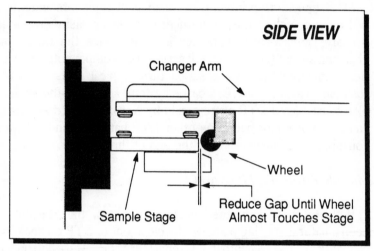

Figure 11. Lining up the wheel with the stage.

13. Looking down at the top of the arm, move the changer to line up the edges of the arm with the sides of the sample stage. This establishes the coarse alignment with the stage. (Sensors in the sample changer refine the alignment automatically when the system is in use.)

14. Tighten the two mounting bolts at the rear of the changer (see step 4).

15. Return the arm to the load position under software control by selecting option *L* from the menu:

OPTION= *L* <RETURN>

16. Select option *E* (exit) to terminate manual mode and return control to the sample changer test menu:

OPTION= *E* <RETURN>

17. The sample changer is now completly installed and ready for operation. Before using it, verify operation as described in Section 2.3.

20

Figure 4. Page from manual produced with DTP technology. Courtesy of Nicolet Instrument Corporation, Madison, Wisconsin.

Table 1.   A Macintosh-Based DTP System

Component	Brand Name	Price
Computer	Macintosh SE (hard disk)	$3700
Word processing program	Microsoft Word 3.1	400
Page layout program	PageMaker 3.0	600
Laser printer	LaserWriter II NT	4600
Scanner	Thunderscanner	250
Display	(Standard 9″)	0
	TOTAL	**$9550**

Table 2.   An IBM-Based DTP System

Component	Brand Name	Price
Computer hardware	IBM PS/2 Model 50	$3600
Word processing program	Wordstar 4.0	400
Page layout program	PageMaker	900
Laser printer	HP LaserJet Series II	3600
Scanner	Handy Scanner HS-1000	300
Display	IBM 8514	$1550
	TOTAL	**$9950**

A typical IBM-based system might include the components shown in Table 2.

Off-the-Shelf DTP System

Just as electronics companies have learned to market stereo components in easily assembled integrated packages, so have microcomputer manufacturers seen the wisdom in selling integrated, ready-to-use DTP systems complete with software. The solution ensures compatibility of all the components, which is probably a bigger problem in setting up a DTP system than it is in setting up a stereo. It also simplifies customer support problems. Of course, along with the peace of mind comes a loss of flexibility.

For those who are willing to make that trade, several systems have been introduced into the DTP market. They are probably most appropriate for businesses and research teams that can afford the luxury of hiring and training a full-time employee to produce DTP documents, but can't afford a computer system manager to oversee setup of a component system.

One such integrated system is the Xerox Desktop Publishing System. Xerox sells and services all the components in the system: the Viewpoint package of graphics and word processing software, the high-resolution 19-in. video display, the Xerox 4045 Model 50 laser printer/copier, and the Xerox 6085 Professional Computer System. The package lists for around $9200.

Other Word Processing Products

Much of the really helpful new technology in the word processing field does not fit neatly into the above categories. Every problematic aspect of writing from spelling to punctuation to organizing has generated software to ease the writer's burden. We discuss these miscellaneous writer's aids below.

But first a warning. They vary greatly in their effectiveness. Spelling checkers, for example, do exactly what they claim to do—they *check* spelling, rather than correct it, and they do so quickly and accurately. On the other extreme are the punctuation aids, which, barring unforeseen advances in artificial intelligence, won't be much help any time soon.

Spelling Checkers

Most best-selling word processors come with a built-in spelling checker. The program we used to produce this book, Volkswriter Deluxe Plus, has an 80,000-word dictionary stored on the program disk. When a certain command key is pressed, the program checks every word in the document being edited. Any misspelled word will, of course, not be in the program's dictionary—unless the misspelling happens to be a word itself. When the spell checker finds a word it doesn't recognize, the program highlights the word in question, and suggests possible correct spellings. These suggestions are simply words that are phonetically similar; they may or may not be what the writer had in mind. Quite often the word in question is a proper noun (such as **Gisela Kutzbach**) that is spelled properly, but is not a dictionary entry. Even with these unavoidable drawbacks, spelling checkers are a great help to even good spellers because they detect so many typographical errors in sch a short time—like the one three words back.

If you already have a word processing program that doesn't include a spell checker, you can buy a separate spell checker program. The Sensible Speller is a good choice for checking files written with many popular Apple II word processing programs, while Word Proof works with documents written on IBMs and IBM compatibles. Spellswell and MacLightning are spell checkers for Macintosh owners.

Writing Style Analyzers

Computers can check writing structure as well as spelling. Whether they will ever be able to correct writing errors is unclear, but with the help of ingenious style analysis programs, they can already make useful comments on style and readability.

A typical style analysis program is RightWriter, which works with most IBM-compatible word processing files (a similar product for the Apple Macintosh is MacProof, which also checks spelling). It starts by reading through a document and comparing the words it finds against its dictionary. It makes special note of slang, jargon and wordy expressions.

More impressively, it makes some rudimentary grammatical analyses. RightWriter picks out adjectives and adverbs and chastises the writer if it finds too many. It also detects passive voice (well, most of the time), and gives an appropriately rude comment when it does so (see Chapter 3). Finally, it determines sentence length and, according to an algorithm that doesn't quite match standard grammatical analysis, makes a judgment of whether a sentence is simple or complex. From all of the above, it scores a piece of writing on writing strength and readability.

The summary comments it gives can help a writer who occasionally needs a reminder, not an editor. Not all of the advice is accurate, but RightWriter is so thorough in its search for wordy expressions, cliches, and passive voice verbs that it is a useful aid.

Science and Math Word Processing Programs

A sticky problem that engineers and scientists have had with basic word processing programs is their inability to reproduce complex equations. Summation signs, integrals, sub- and superscripts were just too much to expect from what was after all a program to process words.

The problem has never been acutely felt by Apple Macintosh users because special characters can be easily produced with the Macintosh's graphics. Long-suffering IBM PC users have only recently been given a solution. Proof Writer and Volkswriter Scientific are two word processing programs for IBM PCs and compatibles that allow reasonably faithful reproduction of equations and scientific notation on a quality dot-matrix printer. These two programs are harder to learn than basic office- or home-oriented programs, but they provide much more control over spacing, placement, and character size.

Some truly heavy-duty math types are turning to an even more powerful (and even harder to learn) solution: a typesetting program called TEX (pronounced "tech"). Developed by a computer scientist, TEX is a high-powered package that gives the user power over more printing parameters (kerning, leading, and ligatures, for example) than most people might want to worry about. But once mastered, TEX allows an engineer or scientist to produce camera-ready copy of text containing the most complex equations. The technology is obviously perfect for production of abstracts and articles for proceedings. that print contents exactly as they are submitted. It is available for IBM and IBM compatibles (PC-TEX or MicroTEX) and for the Macintosh (MacTEX).

Outliners

Professional programmers are by necessity so conscious of the importance of structure and organization that several have written programs specifically to help writers organize ideas into outlines. Perhaps the best known of these "idea processor" programs is ThinkTank, versions of which are available for the Apple IIe, Apple Macintosh, and the IBM. (Of course, word processing programs can help you write outlines as well. This chapter was written from an outline entered into a word processing file. Each major and minor heading was in turn developed by inserting text underneath it. Several blocks were moved and several more were deleted in the process.)

What outlining programs like Thinktank can do that word processing programs can't is to display just the heading structure and not all of the content of a long outline or document. At any point, a user can give either the COLLAPSE or the EXPAND command. COLLAPSE eliminates the details of a particular section so that you can look at the forest, while EXPAND gives you a closer look at a particular tree.

Record Keepers

If you are annoyed by tedium that encroaches on any phase of your computer-assisted writing, you can probably find a software product written to remove it. The particular problem of tedious record keeping of computer use has been addressed by a program called Logit (for the IBM PC and IBM compatibles). Given the concerns of both the IRS and our customers (for those of you in business), it is handy to have a record of how long a computer is used and for what purpose. Logit keeps track of how long, for example, you spent using a word processing program for a writing project. The program greatly simplifies billing and tax accountability.

Productivity Enhancers

One sort of productivity enhancer is typified by Sidekick (for the IBM and IBM compatibles and for the Macintosh). It was created by heavy-duty computer users who wanted to handle all aspects of their work on the computer. Sidekick (and Desqview and Homebase and many others) lets you use your personal computer as a calculator; in another mode it keeps your calendar and reminds you of appointments; in still another mode it acts as a notepad—a handy, computerized place to keep all those notes and reminders that otherwise end up getting lost. Another sort of productivity enhancer comprises those programs that allow you to customize your use of the standard computer keyboard. They let you reassign the keys so that you can custom design a keyboard. They also let you assign a whole word or string of words to a single keystroke. Obviously, your productivity will increase if you can enter your own name or your firm's name with just one keystroke. One such program for the IBM and IBM compatibles is SuperKey.

Communication Helpers

With the help of a telephone and some computer hardware and software, you can communicate with other computer users: co-workers, clients, research associates, or computer-using friends. More important for some engineers and scientists, you can obtain information from one of many commercial services that are in the business of supplying data to computer users.

The computer hardware (what is known as a modem—a "modulator-demodulator") and software needed to communicate with other computers are often sold together. The modem attaches either externally or internally to your computer. When hooked to your telephone, it communicates in a serial (rather than parallel) fashion over standard phone lines with whichever computer (and modem) you would like to reach. The software coordinates things like the automatic dialing, interpretation of the signal on the receiving end, and details of sending and receiving information.

In the IBM world, the best-selling modem package (the Smartmodem 1200) is sold by Hayes. In the Apple Macintosh domain, the Hayes Smartmodem 1200 is also the standard.

Utility Programs

We have neither the space nor the energy to discuss all the miscellaneous programs that relate in one way or another to writing, but we can point you toward a good source of information on them: your local users' groups. Call your Chamber of Commerce or a nearby computer store for information on how to reach one.

Users' groups usually have whole libraries of programs that can be copied by anyone. These programs can be classified either as public domain software, in that the developer has granted free use of the program to anyone, or as "shareware." Developers of programs of the second sort encourage copying and distribution of their programs, but expect anyone who regularly uses one of these programs to contribute $25 to $35. In exchange for the contribution, developers typically send complete documentation and any program upgrades.

Almost all of these programs available from users' groups, whether they be public domain or shareware, do very useful things, but you will never find out about them from your computer dealer because they are not commercial products. An example of a very useful public domain program is PC-Picture by E. Ying. It allows you to create surprisingly sophisticated graphics—and it even comes with a library of ready-made drawings that you can use. Another example is Capslock, a program that keeps you from typing **dEAR sIR.**

Many of these public domain programs can be obtained from what are known as bulletin boards. Usually run as a public service by a users' group or public organization, these electronic bulletin boards are actually computers with modems with which any compatible computer can connect. From these bulletin boards you can obtain messages, programs, or data files which you can send back to your computer's memory, examine, and save if they turn out to be worthwhile. These bulletin boards have become sources for the most current information on new products and problems associated with them. The information that bulletin boards supply is one of the best reasons for buying a modem and communication software.

THE COMPUTER REVOLUTION IN WRITING: A SUMMARY

Computers don't make engineers and scientists more skilled at writing, but they can help make engineers and scientists more productive writers. Personal computers and related technology streamline the writing process and create options for more attractive, more professional-looking documents.

This chapter has described hardware and software that engineers and scientists can use to help them with their writing work. Three levels of hardware and software were described that matched three levels of writers' needs: text, text integrated with graphics, and professional-quality text with graphics. Finally, the array of products available to help engineers and scientists with tasks relating to writing were discussed: outlining software, communication software and hardware, text analysis programs, word processing software especially designed for scientists and engineers, record-keeping software, and utility programs of various types.

We presented all this information knowing that the technology is changing so fast that you will have to consult computer magazines and local users' groups if you hope to keep up to date. Our purpose was to let you know the ways in which computers have changed the writing process for many scientists and engineers. Read from the following bibliography if you would like to know still more.

REFERENCES AND FURTHER READING

Books

Bear, John. 1983. *Computer Wimp*. Berkeley, CA: Ten Speed Press.
Bove, T., C. Rhodes, and W. Thomas. 1986. *The Art of Desktop Publishing*. New York: Bantam Books, Inc.
Lombardi, John V. 1983. *Computer Literacy: The Basic Concepts and Language*. Bloomington, IN: Indiana University Press.
Mali, Paul, and Richard W. Sykes. 1985. *Writing and Word Processing for Engineers and Scientists*. New York: McGraw-Hill Book Company.
Ritvo, Ken, and Greg Kearsley. 1986. *Desktop Publishing*. La Jolla, CA: Park Row Press.
Schwartz, Helen J. 1985. *Interactive Writing: Composing with a Word Processor*. New York: Holt, Rinehart & Winston.

Periodicals

Byte Magazine, McGraw-Hill, Inc., One Phoenix Mill Lane, Peterborough, NH 03458.
COMPUTE! Compute! Publications, Inc., ABC Publishing, 1330 Avenue of the Americas, New York, NY 10019.
InfoWorld, 1060 Marsh Road, Suite C-200, Menlo Park, CA 94025.
The MACazine, Icon Concepts Corporation, 8008 Shoal Creek Blvd., Austin, TX 78758.
MacWEEK, Patch Communications, 5211 S. Washington Ave., Titusville, FL 32780.
PC Magazine, Ziff-Davis Publishing Co., One Park Ave., New York, NY 10016.
PC Publishing Magazine, Hunter Publishing Co., 950 Lee St., Des Plaines, IL 60016.
PC Week, Ziff Communications Company, 800 Boylston St., Boston, MA 02199.
PC World, PCW Communications, Inc., 501 Second St., San Francisco, CA 94107.
Personal Computing, Hayden Publishing, 10 Mulholland Dr., Hasbrouck Heights, NJ 07604.
Personal Publishing, Renegade Publications, Box 390, Itasca, IL 60143.

Trademark Product Names

AutoCad (Autodesk Inc.)
AutoSketch (Autodesk, Inc.)

Bank Street Writer (Broderbund Software)
Capslock (available through electronic bulletin boards)
Desqview (Quarterdeck Office Systems)
Drawing Assistant (IBM)
Filing Assistant (IBM)
Graphics Printer (IBM)
Graphing Assistant (IBM)
Handy Scanner HS-1000 (Diamond Flower Electric Instruments
    Company, Inc.)
Homebase (Software Resource Group)
ImageWriter printer (Apple Computer, Inc.)
Inset (American Programmers Guild, Ltd.)
LaserJet (Hewlett-Packard Co.)
LaserWriter Plus (Apple Computer, Inc.)
Lisa (Apple Computer, Inc.)
Logit (John Beasley, Allan Robertson; Wisconsin Software
    Development Distribution Center)
Lotus 1-2-3 (Lotus Development Corp.)
MacDraw (Apple Computer, Inc.)
Macintosh II (Apple Computer, Inc.)
Macintosh Plus (Apple Computer, Inc.)
Macintosh SE (Apple Computer, Inc.)
MacLightning (Target Software, Inc.)
MacPaint (Apple Computer, Inc.)
MacProof (Automated Language Processing Systems)
MacT$_E$X (FTL Systems)
MacWrite (Apple Computer Inc.)
MicroT$_E$X (Addison-Wesley)
Multimate (Multimate Corp.)
My Word! (TNT Software, Inc.)
OfficeWriter (Office Solutions, Inc.)
PageMaker (Aldus Corp.)
PC-Picture (E. Ying; available through electronic bulletin
    boards)
PC-T$_E$X (Personal T$_E$X, Inc.)
Personal Computer AT (IBM)
Personal Computer System/2 (IBM)
Personal Computer XT (IBM)
PostScript (Adobe Systems, Inc.)
Proof Writer (Image Processing Systems)
Proprinter (IBM)
Publisher (Ventura Software, Inc.)
Reporting Assistant (IBM)

RightWriter (Decisionware, Inc.)
Sensible Speller (Sensible Speller)
Sidekick (Borland International, Inc.)
SmartModem (Hayes Micro Products, Inc.)
Spellswell (Greene, Johnson, Inc.)
SuperKey (Borland International, Inc.)
T$_E$X (American Mathematical Society)
ThinkTank (Living Videotext, Inc.)
ThunderScan (Thunderware, Inc.)
Toshiba P341 printer (Toshiba America, Inc.)
Viewpoint (Xerox Corp.)
Volkswriter Deluxe (Lifetree Software, Inc.)
Volkswriter Deluxe Plus (Lifetree Software, Inc.
Volkswriter Scientific (Lifetree Software, Inc.)
Windows (Microsoft Corp.)
Word (Microsoft Corp.)
Word Proof (IBM)
WordPerfect (WordPerfect Corp.)
Wordstar (Micropro Corp.)
Writing Assistant (IBM)
Xerox 4045 Model 50 laser printer/copier (Xerox Corp.)
Xerox 6085 Professional Computer System (Xerox Corp.)

Chapter 6

Legal Issues in Writing

We live in the age of the lawsuit. Pick up just about any newspaper and you can find half a dozen articles about legal battles among individuals and corporations that involve potential monetary awards of more than a million dollars. Many of those cases involve the work of engineers and scientists. For just a few well-known examples, think of the lawsuits concerning the drug Thalidomide, the Dalkon Shield contraceptive device, the Three Mile Island nuclear reactor, and the space shuttle *Challenger.* As lawsuits have become more frequent, companies and organizations have become much more concerned about the legal ramifications of all their operations.

This chapter is intended to introduce you to some of the legal issues surrounding on-the-job writing. It is not intended as legal advice: the authors are writers and teachers, not attorneys. All we can do is to alert you to some of the areas in which your writing and the legal system might intersect. The chapter begins with an overview of what sorts of documents may be part of legal proceedings, what kinds of circumstances might result in your writing's being legally important, and what you can do to protect yourself and your organization. It then looks in more detail at two specific areas that frequently involve engineers and scientists: product liability and specifications. Finally the chapter includes a short list of references for more in-depth coverage of these topics.

YOUR WRITING AS A PART OF A LEGAL RECORD

As an engineer or scientist, you probably do not think of yourself as central to your organization's legal representation—that is what the company's lawyers are for. But because you are central to the company's work, and that work is sometimes the subject of legal proceedings, your work, and especially your writing, may in fact be important in a legal battle. Every work product, whether from a manufacturing firm, a construction company, or a team of consultants, has a paper trail behind it. The documents in that trail, from the first memos to the final report, constitute a record of the work and may be important in a legal case. This section looks at your role in building that legal record. First, let's look at what can be part of it.

What Is Discoverable Evidence?

Discovery is the legal term for fact-finding and **evidence** is what is used to prove those facts at a trial. Whether something is discoverable legally means whether the opposition has a right of access to the information sought and can force you to produce it as evidence. The simple answer to the question of what is discoverable in a legal action is "practically everything." To illustrate, here is a short excerpt from the Wisconsin Statutes, which closely follow the federal rules governing discovery:

> Parties may obtain discovery regarding any matter, not privileged, which is relevant to the subject matter involved in the pending action, whether it relates to the claim or defense of the party seeking discovery or to the claim or defense of any other party, including the existence, description, nature, custody, condition and location of any books, documents, or other tangible things and the identity and location of persons having knowledge of any discoverable matter. [Section 804.01(2)(a), Wis. Stats.]

What this means is that if your company is the defendant in, for example, a product liability suit, the plaintiff's attorneys can subpoena every piece of paper relating to the product except those that are privileged, such as correspondence protected by the attorney-client relationship. In other words, everything from a handwritten memo suggesting a new product idea to the owner's manual for the eventual product may end up as exhibits in court.

If some of these documents include information that must be kept confidential to protect a business's ability to compete (secret formulas and the like), the court may make special arrangements to protect confidentiality—but that will not keep those documents from being discovered and/or admitted as evidence at a trial.

What Is the Role of the Engineer or Scientist?

Companies and organizations become involved in lawsuits for many reasons. Three legal areas that commonly involve the work of engineers and scientists are patents, product liability, and contract disputes involving construction specifications. In all of these, as an engineer or scientist, you may be deposed (asked to give a sworn statement) or called to testify concerning the technical aspects of issues involved. In addition, documents you have written may have to be turned over to the party who is suing your company through what is legally called a "Request for Production of Documents," (part of the discovery process). Those documents produced may be used as exhibits, that is, evidence, at trial. You may find yourself questioned at length at your deposition about the meaning of what you wrote and then again if you are called as a witness to testify at a trial.

It is beyond the scope of this book to go into any of these areas in much depth, although this chapter does look more specifically at aspects of product liability and specifications writing. In all three areas, however, the role of the engineer or scientist *as writer* is similar. Whether you are writing the description of a mechanism for a patent application, the operating instructions for a new product, or the specifications for a construction project, your job is to be as clear, precise, and complete as you can—if you haven't been, you can bet a skillful lawyer is going to make you (and your employer) wish you had said exactly what you meant!

The other chapters of this book, especially Chapter 3, Writing and Revising the Rough Draft, can help you in general to keep your writing as sharp as possible. In the remainder of this chapter we will look more specifically at writing issues involved with product liability and specifications.

PRODUCT LIABILITY

Since the rise of the consumer movement in the 1960s, people have expected manufacturers to produce safe products—and have been increasingly ready to sue if they were injured by an unsafe one.

In this section, we look at product liability from the point of view of how it affects the writing an engineer or scientist may do. We begin with a brief look at the legal basis for product liability suits, cautioning again that we are not attorneys: a complete exposition of the subject is beyond both our expertise and the scope of this book.

Legal Background

Product liability litigation may be based on any of several legal theories of liability, including negligence, breach of warranty, and strict liability in tort. With any of these, the manufacturer is not liable unless all of the following are true:

- The product has a defect.
- The defect was present when the product left the manufacturer's control.
- Injury or damage occurred.
- The defect caused the injury.

If, for example, a company made an electric drill that after two minutes of normal use became hot enough to burn the user's skin, that manufacturer would probably be liable for burns incurred by the user. But if instead, the user broke a toe when he dropped the drill on it, the manufacturer would probably not be liable, because even though the product had a defect (overheating), that defect did not cause the injury (unless the user dropped it because it was too hot to hold!).

A product can be defective in many ways, including design, marketing, packaging, and so on. For engineers and scientists, the two areas of most concern are the design of a product and the instructions and warnings that accompany it, because these are the areas in which the engineer or scientist plays a key role.

Engineers and scientists are closely involved in the development of product designs. In addition to requiring direct testimony, the parties to a lawsuit may subpoena all sorts of documents that surrounded the evolution of the design: such items as test reports, in-house evaluations, and memos discussing problems with the design may all appear in court as evidence. As an engineer or scientist, you may write some of these documents.

Engineers and scientists are also usually closely involved in the development of the instructions and warnings that go with a product. Even if you don't actually do the writing yourself, you will have to work closely with the technical writer. After all, if you designed the product, you know it better than anyone else and are best qualified to tell the user how to operate it and what hazards to watch out for.

Guidelines for Design-Related Writing

The writing that accompanies development of a design is so varied that it is difficult to develop specific guidelines for it. Typically, this category of writing will encompass a range of materials from proposals to spec sheets and from memos to formal test reports. What all of this material has in common is that it relates to product design and may show up in court as evidence in a product liability suit. You never know in advance, of course, whether these documents will be used in court, but if they are, they may be damaging, especially if they were not carefully written.

The most harmful documents are those that indicate that a manufacturer was aware of a product hazard and did nothing to fix it. An example is the tractor manufacturer that was sued because a farmer was severely burned when he took the gas cap off the tractor and gasoline geysered out of the filler pipe and ignited.

During the course of the trial, documents subpoenaed and introduced as evidence against the manufacturer showed that it had known of this problem for several decades, but had decided on financial grounds to settle personal injury claims rather than change the design. The resulting jury damage award of several million dollars put a halt to that company policy.

Such policy decisions are not usually made by engineers and scientists, although they may contribute to them. As an engineer or scientist, you will more often be involved in writing specifications, test reports, product evaluations, and the like. You may also discover problems with a product or come up with ideas for improving it and communicate these to management in correspondence. With these sorts of documents, your responsibility is to write exactly what you mean and to know what you've said. In other words, the two major guidelines are these:

- **Write precisely.**
- **File carefully.**

Both are important.

For design-related writing, the main guideline is to be precise. A test report, for example, that shows up as evidence in a court case will be less damaging if it is specific about the tests performed and quantitative in its discussion of results than if it uses vague terms like "performed well under normal conditions." A good lawyer can take that sort of imprecise statement and make it sound as though you were attempting to hide something. This does not mean that a summary statement is inappropriate—only that the information must be there in specific terms as well.

Being able to find what you have written is just as important as writing carefully. A thorough and accessible file system is essential. You must be able to produce all the documents relating to a product design when needed. If you cannot find some of them, you appear at best to be disorganized and at worst to be deceptive.

These two guidelines are good advice in general—but they are vital in a product liability lawsuit situation. Equally vital is the care put into writing the instructions and warnings that accompany a product to the end user, and it is that subject to which we now turn.

Guidelines for Writing Instructions

The adequacy of instructions for a product frequently becomes an issue in product liability litigation. A manufacturer may find itself held liable if a user was injured because the operator's manual did not clearly explain how to use the product properly. As a consequence of the increase of litigation, manufacturers are paying more attention to producing good instructions for their products and often hire professional technical writers to produce them. Even so, the engineer or scientist who designed the product still has a major part to play in providing information to the writers and checking their work for technical accuracy. In this section, we look at what the criteria are for good instructions and offer specific suggestions for writing them. For more detailed information on writing instructions, see Schoff and Robinson, *Writing and Designing Operator Manuals* (Schoff and Robinson 1985).

Good instructions tell a user how to use a product properly and safely. They are clear and easy to understand, and they provide all the information the user needs to operate and care for the product. That sounds simple enough, yet we have all struggled with, puzzled over, and fumed at poor operating instructions. Why is it so hard to produce clear, straightforward instructions? Let's look first at what can go wrong.

The most common flaws in instructions fall into the following three categories:

- The instructions assume too much.
- The language is ambiguous.
- The structure is confusing.

Other problems occur as well; for example, the instructions may seem not to apply to your product—which probably means that there was a last-minute design change and the manufacturer didn't want to wait for the manual to be rewritten. But that is not so much a flaw in the writing as a flaw in company policy. The three categories listed cover most of the writing errors.

Instructions Assume Too Much

The first type of flawed instructions includes those that assume too much knowledge on the part of the reader. A good everyday example of these is the instructions that accompany sewing patterns. Unless you already know how to sew, you are out of luck. Directions that tell you to "make a felled seam" or "pink the edges, except for the selvedge" are incomprehensible if you don't know what the terms mean. Similarly, the instruction in a computer manual, "Now enter any valid flag designations, in any order, with or without spaces" means little if you are unfamiliar with computers.

People write instructions that assume too much knowledge because they fail to think about who their audience is or because they suffer from "shop blindness"—they know so much about the product that they can't imagine anyone could be unfamiliar with it. As an engineer or scientist involved with writing an instruction manual, try to think about who will be reading it. Remember that the person who reads your instructions is usually a person who does not already know how to use the product. Even if your product is highly specialized—medical operating room equipment, for example—at least some of the persons reading the instructions will be novices. If your product is one aimed at the general-public market, many of your users will be first-timers. Try to look at your instructions as an outsider; or better yet, try them out on an outsider. If you have to use specialized terms, define them early on, unless you are sure that your readers will know them.

Language Is Ambiguous

The second type of flawed instructions includes those that use ambiguous language. This means simply that the meaning isn't clear. For example, the instruction, "Turn the knob to the right" means different things, depending on whether you think of the top or the bottom of the knob moving to the right. Or take another example, "The pressure should be adjusted to 250 p.s.i." Does that mean check to see if it is adjusted to that pressure or make the adjustment? Ambiguous language makes the reader decide what it means. This is both frustrating for the reader and potentially dangerous, if the reader makes the wrong choice. For your company, it may mean unnecessary costs incurred through settling or defending a product liability claim.

Usually, ambiguous language results either from carelessness or lack of awareness of different possible interpretations. This flaw is especially difficult to correct alone, because you know what you meant when you wrote an instruction: you have the image correct in your mind, and it is hard to look at the words you have written from the outside to see what they really say. A well-prepared lawyer is going to challenge you on exactly this point: why didn't you say exactly what you meant? Your attempt to clarify will undoubtedly make the lawyer's point, and further attempts to clarify may make the situation worse. A valuable practice is always to have somebody else read through your instructions to make sure they're clear. Take your reviewer's criticisms to heart and make the necessary corrections.

Structure Is Confusing

The third type of flawed instructions includes those that are structured in a confusing way. They may be confusing because the directions are not given in the order they should be followed, or they are hard to distinguish from background material, or they give the reader too much detail before explaining the overall process. Whatever the form of the confusion, the result is that the reader is left to puzzle out the meaning. Again, the result may be that the reader makes an incorrect choice about what to do.

Following a few tricks of the trade will help you to organize instructions clearly. (Most of these "tricks" are dealt with more fully elsewhere in this book, because they are also general principles of good writing). To make sure your instructions are organized clearly, follow these basic rules:

- Give instructions in the order you want them followed.
- Clearly separate instructions from other material.
- Give general information before specific.
- Put instructions in parallel form.

The reason to give instructions in the order you want them followed seems obvious, and yet surprisingly often instructions are not presented in that manner. Instead, this sort of list will appear:

1. Be sure surface is clean and dry and free from dust.
2. Apply adhesive to both surfaces to be joined.
3. Let adhesive dry for 5–10 minutes or until it is tacky.
4. Firmly press surfaces together, clamping if necessary to maintain pressure.
5. Allow to set until fully cured (24 hours).
   Note: solvents present in adhesive may damage some plastics and vinyl tile. Before using on such surfaces, test adhesive on a sample to ensure the surface is suitable.

By the time the user gets to the cautionary note, the damage may already be done. You can avoid such damage by thinking about the user's process in reading the instructions and using the product. The truth is that most people do not read through the whole set of instructions before starting a procedure—even if the instructions tell them to. The writer therefore has to be certain that when a user reads a step, he or she already has all the necessary information to do that step properly.

Following the second rule, "Clearly separate instructions from other material," makes it easy for the user to keep on task and not lose his or her place. Most times, someone reading instructions is looking back and forth from the instructions to the product, referring to the directions while performing a task or procedure. Physically separating instructions from other material, such as product description, theory of operation, specifications, and so on, makes it harder to get lost. A good way to separate instructions from other material is to put them in a list format. That way you can physically set them off from other material with white space.

The third basic rule is to present general material before specific. This is good advice for writing overall, and Chapter 2, Effective Organization, goes into more detail on this concept. Using a general-to-specific order is especially important in writing instructions because so much of the material is, by definition, very specific and detailed. Unless you orient the reader first by providing the "big picture," all those specifics will seem very disjointed and fractional. Sometimes the best way to do this is to provide a summary of the procedure at the beginning and then give the step-by-step instructions. Another way is to segment a long process into sub-procedures, each with four or five steps. Both of these presentation techniques help the user to see how each small step fits into the procedure as a whole. Knowing the context makes it much easier to understand a particular detail.

The final rule is to put instructions in parallel form. This is also addressed elsewhere in the book (see Chapter 2, pp. 37–39 and Chapter 3, pp. 67–68), but deserves brief mention here. Especially if you use a list, be sure that your instructions are written so that they are grammatically parallel. That way, once the user sees the first instruction, he or she knows the pattern for all the subsequent instructions and can concentrate on the what information is being conveyed rather than on how the sentences work. For instructions, the best pattern is to start with a command-form verb:

- **Remove** the snap ring.
- **Enter** the data.
- **Clean** the spindle.
- **Measure** the gap.
- **Calibrate** the instrument.

Instructions are actually easy to write well, if you take the time to think about them from the end-user's point of view rather than from the designer's. A good way to test how well you have done is to watch a novice user (your basic "person-on-the-street") try to operate the product from your instructions. Noting the parts that puzzle the reader or, worse yet, set the reader to doing the wrong thing, will alert you to the parts you need to clarify. Talking with the reader should help pinpoint the problems in the writing. A good rule of thumb is that if your "person-on-the-street" can't follow the instructions you've written, then a lawyer is more than likely to find a jury that will say those instructions do not meet a "reasonable person standard," (which loosely means that they would not be clear and adequate for a person of common knowledge and skills).

Clear instructions are vital to help protect your company from lawsuits resulting from improper use of your products. Equally important are the warnings that alert users to hazards inherent in the product.

Guidelines for Writing Warnings

Failure to warn properly of risks and hazards inherent in a product may be considered a defect, just as inadequate instructions can be. In this section we look briefly at how to develop good warnings. First, we will look at the manufacturer's legal responsibility, then at what constitutes a good warning, and finally, at guidelines for developing warnings.

The Duty to Warn

The basic legal obligation imposed by the duty to warn concept is that a manufacturer has the responsibility to warn against any risk or hazard associated with its product, unless that risk or hazard is obvious. For example, you don't have to warn that if a user hits a thumb with a hammer, it will hurt. Hammers are intended to drive nails, and the danger of smashing a finger is obvious. However, most hammers now do carry a warning against hitting hardened steel items (such as another hammer) because of the not-so-obvious danger that the blow might chip off metal fragments and send them flying. That danger is not obvious. The manufacturer must warn of the hazard even though it results from using the hammer in an abnormal way. The duty to warn includes hazards inherent in foreseeable misuse, as well those encountered in normal use.

A manufacturer must warn anyone who might reasonably come in contact with the product, even if that person is not the expected user. This means that just as you must analyze your audience when you write instructions, you must analyze the user when you develop warnings. Remember when defining your audience for warnings that we live in a pluralistic society and one in which many barriers have fallen in the last few decades. Many products that used to be restricted largely to "professional" users now are widely available to all sorts of people, ranging from novice to experienced. For example, think about the change over the last twenty years in the potential operators of these products: computers, microwave ovens, chain saws, gasoline pumps, and hair dryers.

In addition to a user's degree of experience, another characteristic that is less predictable nowadays than it used to be is familiarity with English. The United States is experiencing another wave of immigration, which means that manufacturers cannot count on their users being able to read warnings printed in English. And of course, there is the continuing problem of illiteracy: even if an operator speaks English, that does not mean that he or she can read it. As a consequence, manufacturers have had to develop innovative strategies to reach users with appropriate warnings.

Criteria for Good Warnings

The courts have generally agreed that a good warning has these characteristics:

- It tells the gravity of the risk.
- It explains the nature of the risk.
- It tells how to avoid the risk.
- It is clearly communicated to the user.

As an example of a *bad* warning, consider this one:

> The solvents in this product may be hazardous to your health. Use with adequate ventilation and avoid prolonged contact with the skin or prolonged inhalation of the vapors.

If you compare it to the listed criteria, it fails on all counts. It does not tell the gravity of the risk—**hazardous to your health** could mean anything from "giving you a headache" to "killing you." It does not explain the nature of the risk: Does it make you cough? Is it a nerve poison? Does it blister your skin? It does not tell how to avoid the risk—it just tells you to avoid it. It is not clearly communicated, in part because it is too vague and in part because it uses fairly high-level language, such as **inhalation.**

By contrast, this bleach bottle warning is much clearer:

WARNING: This product contains chlorine. Mixing this product with other household cleansers, such as toilet bowl cleaners, rust removers, and products containing ammonia, may release deadly chlorine gas. Do not use this product with any other chemicals.

This warning tells the gravity of the risk (deadly), the nature of the risk (chlorine gas), it tells how to avoid the risk (don't mix the product with other chemicals), and it is clearly communicated (it uses everyday language).

Suggestions for Developing Warnings

As an engineer or scientist developing a product, you will be closely involved in identifying the hazards that need to be warned against and in developing the appropriate warnings for the product and the manual. You should of course work closely with your company's legal department, but the following are some general suggestions for making your warnings effective.

- Use simple, straightforward, down-to-earth language.
- Never mix warnings and instructions.
- Follow any standards or practices in your own industry.
- Develop a good format and follow it consistently.

Let's look at these in more detail.

First, use *simple language.* Don't try to "dress up" the warning with pretentious language or make it less scary with clinical terms. In other words, don't say "may be hazardous" when you mean "will result in death." Don't say "can result in bodily harm" when you mean "can amputate fingers." Say what you have to say as clearly and simply as you can. Don't worry that potential buyers might be frightened off by clear language—they are more likely to appreciate the straightforward warning.

Second, *avoid mixing warnings and instructions.* Obviously, if you put warnings in the operator's manual, they will appear together with the instructions, but they should stand out as different from ordinary instructions. You can use color, special symbols, and layout to highlight the warnings. However you do it, make the words you use show up clearly as warnings. If you integrate warnings with text it is too easy for a reader to skip over them. The object is to make a warning hard to miss, so that the user will see it even if he or she does not read the entire user manual word for word.

Third, *follow existing standards and practices* in your own industry. If the American National Standards Institute (ANSI) or another agency or association publishes standards for warnings in your area, you must comply with them. However, meeting the standards alone may not be enough, since the courts may treat them as minimum requirements.

Finally, *develop a good format for warnings* and follow it consistently. Presenting your warnings in a consistent format automatically alerts your readers to the presence of a warning. One effective format that is gaining wide acceptance consists of three parts: a signal word and color to indicate the degree of hazard, a pictogram to show the type of hazard, and a verbal message to explain the hazard and tell how to avoid it. Figure 1 shows an example of a warning label using this format. The signal words and colors most often used are these:

DANGER (red):	probable severe injury or death
WARNING (orange):	possible severe injury or death
CAUTION (yellow):	probable minor injury or product damage

Westinghouse and the FMC Corporation both publish detailed guides for creating warning labels using this format. These can be purchased from the companies for a nominal cost.

Good warnings will help to protect your product's users from injury and your company from product liability lawsuits. The guidelines presented here outline some of the key aspects of effective warnings that you as an engineer or scientist should consider and apply to your company's products.

A legal area that is different from product liability but related to it is liability resulting from disputes over construction specifications. The last part of this chapter looks at specifications development and how the engineer can apply principles of good writing to minimize legal problems in that area.

Figure 1.   Example of a warning label. The field surrounding the word *danger* would ordinarily be red. Printed with permission from FMC Corporation.

CONSTRUCTION SPECIFICATIONS

Any construction project is the result of cooperation among several parties: the owner, the engineer, scientist, or other design professional who designs the project, and the contractor who builds it. Sometimes disputes arise among these parties when the completed project is not satisfactory to the owner, the work is not performed in the way that the owner or design professional thinks it should be, or the contractor finds it impossible to meet the conditions called for by the designer. In such a dispute one or the other of the parties may be found liable for damages. Often, the dispute centers on interpretation of the requirements of the contract and the specifications on which it is based. In this section we will look at your role as engineer or scientist in developing specifications and offer suggestions for writing specifications to minimize problems brought about by lack of clarity. Unclear wording is just one of a number of potential pitfalls in specifications writing, but the scope of this book limits our treatment to this area.

We begin with some background on the bidding process and how specifications, contracts, and change orders fit in, then look at typical problem areas, and finally present some guidelines for writing good specifications.

Background

When a construction project is undertaken, the parties to the project usually enter into a contract for the work to be done. The overall process by which the contract comes about is called the bidding process. This section looks at that process, including the specifications that are part of it, the contract that results, and at modifications, or change orders, that may be required before the project is completed.

The Bidding Process

The bidding process varies depending on who is requesting work to be done—and the role of specifications varies as well. The two basic kinds of situations are when a *private individual* contracts for work and when a *public entity*, such as a city or another unit of government, contracts for work. Both situations usually call for specifications to be written, but the role of the engineer or other design professional may vary.

If an individual wants, for example, a private sewage disposal system installed on a farm, he or she might come to you (as engineer) and request that you design it. Your responsibilities might include not only design but also construction, so that you might also serve as general contractor. In public construction, the situation is usually quite different.

Most public construction projects are subject to competitive bidding laws. If a city, for example, wants to build a wastewater treatment plant, the governing state statutes typically require it to develop specifications and publish them in a request for bids. Contractors then bid on the project. Competitive bidding laws then frequently require the city to write very detailed specifications, because allowing contractors to write their own undermines the competitiveness of the bidding process.

Specifications

The specifications that an engineer or other design professional prepares become the basis for pricing the work and a standard for judging the contractor's work product. The different contractors bidding on the wastewater treatment plant in the example mentioned above would all use the same set of specifications to figure their bids. Specifications therefore must be sufficiently detailed to enable the contractor to calculate the project's cost and sufficiently reliable to ensure that the calculated cost is accurate. If they are not detailed and reliable, disputes between owner and contractor are practically inevitable.

Specifications may suffer from a variety of ills, including technical inaccuracy, but here we are concerned only with problems in the writing. Writing problems in specifications come in a variety of forms, but the basic characteristic of all is *lack of clarity*. If the specifications are not clear, they admit more than one interpretation, and differing interpretations lead to disagreements. The law refers to contracts that contain such language as ambiguous and consequently subject to judicial construction (that is, either a judge or a jury is going to decide what the words used in the contract mean). Your task as engineer or designer is to write your specifications with clarity and precision. Of course, that is your task as technical writer in any situation, but it is especially important with specifications because they form the basis for the contract.

Contracts

A contract is a legally enforceable agreement. To be enforceable, the actions the parties promise to perform must either be clearly expressed in the contract or ascertainable. For a construction contract, such as for the wastewater treatment plant discussed above, the specifications form the basis of those promised actions. If the contractor builds a plant in any way different from the way prescribed in the specifications, the contractor may be found to have breached the contract, unless some clause in the contract permitted alterations.

Contracts frequently contain provisions for adjustments of the work or payment to reflect changing circumstances. Common examples are escalator clauses, which allow an increase in payment to the contractor if inflation drives prices up during the term of the contract, or excusable delay causes, which excuse a contractor for delays that are out of its control, such as those caused by a strike.

One type of clause that relates directly to specifications writing is the "changed conditions" clause, which covers problems arising from unforeseen construction conditions. These might include such things as the contractor's encountering rock in an excavation where initial site investigations indicated only loose material. The contract, based on expected conditions, would not otherwise provide for compensation for the increased difficulty of excavating through rock. Changed conditions clauses have the effect of providing for equitable sharing of the risk of unforeseen conditions among the parties to a contract. For the engineer or design professional, they acknowledge the fact that however careful you are in your investigations and your specifications writing, you cannot know everything.

Whether or not a contract contains a changed conditions clause, any departure from the original contract requires a change order that, in effect, amends the contract. Typically, this change order will not be effective (contractually binding) unless it is in writing and agreed to by both parties to the contract.

Change Orders

The need for changes arises from many different sources: unforeseen conditions, design deficiencies, or changes in the owner's needs (especially in long-term projects), to name a few. One important source of modifications is ambiguity in the specifications. If the specifications are unclear or subject to interpretation, a change order may be needed to clarify exactly what is required. Without such formal modification, disputes can arise when a contractor, interpreting the specifications one way, proceeds with work that the owner, interpreting the specifications differently, sees as a change to the original contract. Again, the need for clear and unambiguous specifications is obvious. The next section looks at some of the problem areas in specifications writing that may lead to ambiguity.

Problem Areas in Specifications Writing

Problems arise when the wording of specifications is ambiguous or unclear. The general comments elsewhere in this book regarding clarity certainly apply to writing specifications. However, specifications have certain problem areas of their own. This section looks at these, in particular, the following:

- "or equal" specifications
- mixed design and performance standards
- ambiguous or confusing wording

"Or Equal" Specifications

Often construction specifications will call for a particular brand-name product "or equal" product. The difficulty here lies in the term **or equal**. The term **equal** is too vague to settle a dispute over the suitability of a given product because it does not specify in what respect the product must be equal to the brand-name one. As a specifications writer, you should always clarify the standards that an alternative product must meet. State the needed qualities in terms that can be quantified: equal compressive strength or corrosion resistance, for example. Be as precise in your requirements as you can.

Mixed Design and Performance Standards

Design standards specify the materials and methods the contractor must use to complete a given project. Performance standards specify the desired result. Mixing the two in a set of specifications may cause a dispute. For example, a set of specifications might call for building a refuse incinerator of certain dimensions, but also state that it should be capable of burning so many tons of refuse per day. The contractor could follow the design specifications to the letter, but the resulting incinerator might have less than the required capacity.

In such a case, who is responsible? The contractor, who did, after all, perform the work required in the specifications, or the designer? The courts have held that specifications carry with them an implied warranty of reliability, and, consequently, the designer would be responsible (Natkins, Smith, and Swearingen 1985, 122). The way to avoid such problems is to be very clear about which way you want to specify the work and not to mix the two. Define your choice if you've made one; otherwise, whenever you offer different approaches to a definition, you open the door to different interpretations.

Ambiguous or Confusing Wording

Disputes most often arise over unclear wording. For example, consider a specification calling for "three foot long spacers" between pipes. Does this mean three spacers, each *one* foot long or an inde-terminate number of spacers, each *three* feet long? The absence of a hyphen or two makes this specification open to different interpreta-tions. Or consider again the refuse incinerator referred to above. When the specifications call for it to handle so many tons of refuse per day, does that mean per 24 hours or per 8-hour work shift? Wording that is not precise, or specifications that are not measurable (that the work should be performed in a "good and workmanlike manner," for example) may become the focus of differences in a protracted lawsuit.

One of the chief difficulties in eliminating ambiguities from your writing is that when you write a specification you know what you mean—and it is not always easy to recognize where an ambiguity lies. It is relatively easy to recognize a phrase like "a sufficient number" as being open to interpretation (unless you specify sufficient to do *what*), but it is not so easy to see a specification calling for so many tons per day as ambiguous.

Guidelines for Writing Specifications

Specification writing is a vital part of the bid preparation process, but too often it takes second place to preparing the plans and other bidding documents. Design companies frequently view the specifications as taking too much time or costing too much to produce, and so shortchange their preparation. Yet in the long run, the extra investment in producing correct and clear specifications will pay off handsomely in avoiding unnecessary and costly litigation. In general, taking care to follow the advice in Chapter 3 will go a long way to ensure that your specifications are clear and unambiguous. In addition, the following suggestions relate particularly to specifications:

- Write new specs for each job.
- Quantify tolerances and standards.
- Use words precisely.
- Punctuate correctly.

The following paragraphs explain these in more detail.

Write New Specs for Each Job

When your company produces many bids for similar jobs, it is very tempting to save time and money by copying part of the specifications from previous bids. However, each job is unique, and recycling old specifications without checking for technical accuracy and applicability to the new situation is asking for trouble. A good way to save some time, especially if you have a computer system, is to have a master set of specifications with all the basic clauses. You can then use this as a guide each time and customize the specifications to fit the job. That way you aren't simply repeating old specs; you are instead looking anew at each part and checking to see if it applies. But you are saving money by not having to start from scratch each time.

Quantify Tolerances and Standards

Whenever possible, use numbers to specify acceptable quality rather than words. Numbers are exact and not subject to interpretation. Numbers are measurable. By contrast, words are slippery things, even if you avoid obvious ambiguities. Instead of directing the contractor to apply plaster to an adequate thickness, for example, specify 1/2 inch (or another appropriate dimension). If words result in a dispute, the courts are more likely to hold in favor of the contractor. A basic rule of contract law is that the contract is construed against its drafter, which here typically is the designer, because he or she prepared the specifications on which the contract is based.

Use Words Precisely

Be sure that you have used the proper word and placed it correctly in the sentence. For example, do not use **above** when you mean "more than," because **above** also has a spatial meaning. Do not say "insure" unless you are talking financial obligations; **ensure** means "to make sure of." Make sure also, that your word order reflects your meaning. Consider these two sentences:

1. Contractor is only required to apply two coats of paint.
2. Contractor is required to apply only two coats of paint.

The first sentence relieves the contractor of all responsibilities except painting; the second limits the number of coats to be applied.

Punctuate Correctly

Correct punctuation may seem like a minor point, but it can have major implications in the interpretation of specifications. We have already looked at the matter of hyphens: the difference between three foot-long spacers and three-foot-long spacers, for example. Be sure that you have placed commas correctly to convey the meaning you want. For example, consider the following two requirements:

1. Contractor will stain all siding which is redwood.
2. Contractor will stain all siding, which is redwood.

The difference the comma makes is considerable. In the first sentence, only the portion of the siding that is redwood gets stained. In the second sentence, all the siding gets stained. The last clause is just added information. (Of course, the whole problem could have been avoided by rewriting the sentence to read "Contractor will stain all redwood siding.")

These suggestions cover the most common problems in ambiguous or unclear specifications, but they do not cover the whole topic of specifications writing. Many other factors go into producing a good set of specifications: technical accuracy, site investigation, familiarity with construction methods and materials, and so forth. The actual writing is nevertheless a critical component. Following these guidelines should help you to produce clear and readable specifications and protect your company from disputes over what contractually agreed-to specifications really mean.

SUMMARY

The writing that engineers and scientists do can be legally crucial, as this chapter shows. Your writing constitutes part of the legal record of a project and might end up as evidence in litigation. Thus, part of your job responsibility as an engineer or scientist is to make your writing precise and to keep accurate records of what you've written.

As an engineer or scientist, you are likely to be closely involved with product development, both in design and production and in producing the instructions and warnings that accompany a product. Consequently, you need to know how the potential for product liability lawsuits can affect design-related writing, instructions, and warnings. In all these areas, precision and clarity are vital, and this chapter attempts to provide specific guidelines for writing effective instructions and warnings that will lessen the likelihood of a lawsuit succeeding against your company.

Finally, if, as an engineer or scientist, your job entails writing specifications, particularly construction specifications, you need to know how specifications fit into the bidding process and form the basis of the construction contract, and also how unclear or ambiguous specifications can lead to costly disputes. You need to recognize some of the common problem areas in specification writing, and make sure that your specifications are clear and precise.

This chapter only scratches the surface of the legal aspects of writing. In a book of this size and scope, we could not treat the subject in much depth. What we have done is to point out some of the areas in which the writing of engineers and scientists may have legal implications. We have suggested ways that you can make your writing better—preventative writing, so to speak—so that if it does become important in a legal dispute, it will not be held to have caused the injury or harm suffered by whoever started the lawsuit.

Once again, we wish to emphasize that we are not giving legal advice; we are giving writing advice. We do not guarantee that if you follow the suggestions in this chapter you will win every lawsuit. We do believe, however, that by following our suggestions, you will write more clearly and precisely, and that in itself may keep you out of some legal disputes altogether.

REFERENCES

Natkins, Burt P., Robert J. Smith, and Pamela S. Van Swearingen. 1985. *Public Construction in Wisconsin.* Madison, WI: Local Government Services, Inc.

Product Safety Label Handbook. 1981. Trafford, PA: Westinghouse Electric Corporation, Customer Service Section, Westinghouse Printing Division.

Product Safety Sign and Label System. 1980. 3rd ed. Santa Clara, CA: FMC Corporation, Central Engineering Laboratories.

Schoff, Gretchen H., and Patricia A. Robinson. 1984. *Writing and Designing Operator Manuals.* Belmont, CA: Lifetime Learning Publications. (Available through Van Nostrand Reinhold Company, New York.)

Appendix 1

Common Errors

This section provides a handy reference for avoiding the most common grammar and punctuation errors found in technical writing. It is not a complete English handbook: many of those are available, some longer than this entire book. Instead, this appendix calls your attention to a few of the mistakes that appear frequently, and helps you to keep them out of your writing.

Grammatical Errors

1. Agreement

English requires agreement between subject and verb and between pronoun and antecedent (the noun to which a pronoun refers). *Agreement* simply means picking the right form of verb or pronoun to go with a given subject or antecedent.

Subjects and verbs must agree in number and person. *Number* refers to whether the subject is singular or plural and *person* to whether the subject is speaking, spoken to, or spoken about:

She was (not *were*) hired (singular).
I am the director (first person).
You are the director (second person).
He is the director (third person).

Few engineers or scientists would be likely to write "She were hired," yet sometimes choosing the right form is tricky. Here are rules for three especially confusing situations.

153

- **The subject is modified by a phrase with plural words.**

    The case of instruments *was* (not *were*) in the truck. (The subject is **case**. **Of instruments** is just a phrase that modifies the subject.)

- **The subject is joined to another noun by something other than "and."**

    The report and the recommendation *were* presented at Friday's meeting.
    The report, together with the recommendation, *was* (not *were*) presented at Friday's meeting. (Two words joined by **and** form a plural subject. Two words joined by other expressions do not. The first word, **report**, remains a singular subject.)

- **One singular and one plural word are joined by "either/or" and "neither/nor."**

    Neither the salary nor the benefits *are* satisfactory.
    Neither the benefits nor the salary *is* satisfactory. (The word nearer the verb governs it.)

Most other subject-verb questions can easily be resolved by remembering that plural subjects and compound subjects (more than one noun joined by **and**) take plural verbs, and singular subjects take singular verbs.

Another kind of agreement is that between a pronoun and its antecedent. The pronoun must agree in *number, gender,* and *person* with the noun to which it refers. Thus, when writing about a particular individual, John Smith, we say *he* is a physicist (singular, masculine, third person). Matching up pronouns is normally no problem. However, English lacks a singular, third person pronoun appropriate to refer to humans that is not gender-specific:

The dog ate *its* dinner (third person, singular, neuter).
The engineer ate *his* or *her* dinner (third person, singular, masculine or feminine).

It is incorrect to use **their** in place of **his** or **her** when the antecedent is singular. Use of **his** to cover both men and women is traditionally correct, but sexist. The best solution is to recast the problem sentence in the plural or in a way that does not require a pronoun:

The engineers ate their dinners.
The engineer ate dinner.

2. Case

Case is a change in the form of a noun or pronoun to show whether it is acting as a subject, object, or possessive in a particular sentence. Thus, we write "*I* wrote the report," but "The results came to *me*," even though both pronouns refer to the same person. In the first instance, the pronoun is the subject of the sentence, and in the second instance, it is the object of the preposition **to**.

In English, nouns do not change their form to show case except in the possessive, where **scientist**, for example, becomes **scientist's**. Pronouns, however, do change form to show all three cases, subject, object, and possessive:

Subject:	*Object*:	*Possessive*:
I	me	my
you	you	your
he, she, it	him, her, it	his, her, its
we	us	our
you	you	your
they	them	their
who	whom	whose

As with agreement, most of the time we have no difficulty choosing the right form, but some situations are apt to be confusing.

- **Use of "me" in formal writing**. Many of us were taught to avoid using the word **me** in formal prose—and left with the impression that it was somehow slangy. Two "rules" are probably at work here: the belief that first person is inappropriate in formal writing and early correction for improper use of **me** as subject (me and Joey went to the lake). In reality, it is perfectly correct to use first person (I, we, me, us) in formal writing—and usually a lot better than convoluted attempts to avoid it. **Me** is also perfectly good English—and in fact is the *only* correct form for the first person singular object case:

If you have questions, contact Jim or *me* (not *myself*). (**Me** is the object of the verb.)

- **Use of who and whom.** Once you can recognize subjects and objects, it is easy to tell when to use **who** and **whom**. Remember that with a dependent clause, it is the function of the pronoun within the clause that governs case, not the function of the clause.

Who is going? (subject of sentence)
Whom should I talk to? (object of preposition **to**)
Tell *whoever* answers that we're ready. (**Whoever** is the subject of the verb **answers**; the clause **whoever answers** is the object of the verb **tell**.

- **Spelling of "its."** The possessive pronoun has no apostrophe. **It's** is a contraction for **it is**.

- **Spelling of other possessives.** Ordinary nouns do use an apostrophe to form the possessive. All singular nouns and plural nouns that do not end in **s**, such as **women**, form the plural possessive with **'s**. Plural nouns ending in **s** just add the apostrophe:

  one woman's calculator
  two women's calculators
  two engineers' calculators

Sometimes people confuse the plural with the possessive, especially for words ending in **y**.

  the company's policy (not companies)
  the companies' policies (more than one company)

3. Misplaced and dangling modifiers

The problem of misplaced and dangling modifiers was addressed in Chapter 3 (pp. 64–65) as an obstacle to easy understanding of material. Misused modifiers may also lead to misinterpretation or provide unintentional amusement to the reader. The cure, of course is to make sure that the noun or verb being modified is present in the sentence and to place the modifier close to it.

Being confined to a small cage, the zoologist tried to enrich the baby
gorilla's environment. (The zoologist is in the cage.)
I looked for the journal article about toxic mold on the shelf in my
office. (The office needs cleaning.)

Dangling modifiers often show up in sentences with passive-voice
verbs—because the actor has disappeared from the sentence:

Taking a deep breath, the surgeon began the incision (active).
Taking a deep breath, the incision was begun (passive).

PUNCTUATION ERRORS

1. Semicolons

Semicolons are used in sentences in two ways: to separate two
independent clauses without using a conjunction and to separate
members of a series that themselves contain commas.

John wrote the report; Susan provided the data.
John wrote a well-planned, organized report; a persuasive, incisive
cover letter; and an effective, creative action plan.

Semicolons are *not* used

- to introduce a list
- to follow the salutation in a letter
- to set off a dependent clause or introductory phrase

2. Colons

Colons are used to introduce lists or explanatory material when
preceded by an independent clause and to follow the salutation in a
letter.

To operate the control panel, follow these steps:
The most important aspect is the simplest: cost.

Colons are *not* used between verb and object or verb and subjective complement, even if the object or subjective complement contains several items.

> The design criteria are size, strength, and weight. (No punctuation after **are**.)
> You must check the fuel, the air supply, and the vacuum pressure. (No punctuation after **check**.)

3. Commas

Commas have more different uses than any other punctuation mark in English, and engineers and scientists use them correctly most of the time. Two errors, however, often appear: using a comma (instead of a semicolon) to join independent clauses without a conjunction and using a comma to indicate where the writer would pause in speaking a sentence, but where there is no grammatical reason for punctuation.

> We are developing a plan of action, it will be ready tomorrow. (Wrong. Two independent clauses require a semicolon or a comma and conjunction to join them.)
> We have developed the plans, and specifications. (Wrong. No need to punctuate between compound object.)

4. Hyphens

Engineers and scientists rarely put hyphens in the wrong places, but they frequently omit them when they are needed.

Hyphens are required to join compound adjectives that precede the noun they modify and to join adverb-adjective combinations that precede the noun:

> The high-strength steel is preferable.
> We need a digital-readout display.

Hyphens are not used between adverb-adjective combinations when the adverb ends in **ly**:

> This is a highly profitable alternative.

Hyphens are also used to join numbers and their units when they are used before a noun, especially to avoid confusion:

The specifications require three-foot-long supports.
The specifications require three foot-long supports.
We requested 12 10-digit codes. (Alternatively, twelve 10-digit codes.)

This short summary covers briefly the most frequent grammar and punctuation errors in technical writing. It is by no means a comprehensive grammar handbook, but it should serve as a quick reference for common trouble spots.

Appendix 2

Equations and Abbreviations

Try to follow the conventions established by engineers and scientists for presenting equations and handling abbreviations. This short appendix covers only the most general situations. If you use many equations and abbreviations, consult the technical style manuals mentioned at the end of Appendix 3.

CONVENTIONS GOVERNING ABBREVIATIONS

Abbreviations can improve technical writing when used judiciously. They are not discouraged in technical writing as they sometimes are in other types of writing. Use them to abbreviate units of measure, to streamline discourse by shortening words or phrases used repeatedly, and to replace very common foreign words or titles.

Abbreviating Units of Measure

Most engineering and scientific journals use standard abbreviations of the units in the SI (initials of the French words for International Scientific) system of measurement. Some measurements and units are shown below:

13.68 m/s
2.62 kg
117.4 a/s

Do not use periods after the abbreviations, and do not make the abbreviations plural. However, for some non-SI abbreviations (such as for inches) use periods to prevent abbreviations from being misread as words:

11.44 in.
Fig. 6.4

Abbreviating Common Words and Phrases

Use abbreviations for those common words and phrases that were derived from Latin, or that have become more common in their abbreviated form. Use periods with these abbreviations, except those that are capital letters, as shown below:

viz. (namely)
e.g. (for example)
i.e. (that is)

but

IQ
PhD
RMS

Abbreviating Frequently Used Phrases

Many times, authors avoid space-eating repetition of key phrases by substituting initials or acronyms. Some of these abbreviations have become part of our everyday language: UFO and IRS, for example. However, before coining a new abbreviation for a phrase in a report or journal article, consider the dangers raised for your readers. Abbreviations and acronyms require considerable effort by your readers, who are probably less familiar with the terms than you. Just as an example, you might be quite used to the term "CD ROM," but many of your readers might have to think hard to bring to mind what you mean (compact-disk read-only memory).

Therefore, we recommend that you coin new acronyms and abbreviations as rarely as possible (one per chapter, article, or report, if possible), and use currently used acronyms sparingly. The first time you use a phrase for which you would like to substitute an acronym, spell out the phrase in full and immediately give the acronym in parentheses, as in the following example:

New guidelines for heavy metal contaminants have been proposed by the Environmental Protection Agency (EPA).

CONVENTIONS GOVERNING EQUATIONS

First make sure that you really want to use equations. They are helpful, even essential, to writing for some purposes and audiences, but counterproductive when used other times and places. Do not assume that because a project required calculation of equations that your project report must too. For example, in most engineering reports, equations are appropriate only in appendices. Similarly, articles for scientific journals should contain equations only if they are part of a derivation, different than the ones usually used, or especially important to the paper's outcome.

If, after analyzing your audience and purpose, you are convinced that your writing should include equations, follow these guidelines:

- Equations that are part of a derivation, or complex calculation, or for some other reason are referred to subsequently in text should be "displayed"; that is, they should be set apart from the text and numbered. Display an equation by centering it horizontally on its own line and vertically with respect to the equal sign. Place its number in square brackets or parentheses flush with the right margin. Define all variables below the equation.
- Other equations, those that can be typed on one line of text, can be included within the text.
- Use a black fine-line felt-tipped pen to add symbols such as Greek letters or radical signs that are not available on your typewriter or printer. If you are submitting an article for publication, make marginal notes as needed to distinguish between easily confused symbols (between upper case chi and X, for example). See one of the style guides listed at the end of Appendix 3 for more information on handling equations that will ultimately be typeset.

- When possible, use the virgule (/) to separate terms of a fraction. Thus, type $y = (x + 4)/m$ rather than $y = \dfrac{x + 4}{m}$.

Appendix 3

Preparing an Article for Publication

"Publish or perish" is the cliche coined to describe the situation of young engineers and scientists in academics. It probably does not overstate the importance of publications in establishing credentials, developing a reputation and ultimately obtaining tenure. Scientists and engineers in industry can also, of course, enhance their careers—and enhance their employers' reputations—by publishing the results of their intellectual efforts. This appendix describes the mechanics of the process of preparing and submitting a manuscript for publication. It assumes that you in fact have information worthy of being in print. Remember that publish *and* perish is possible—the consequence of publishing questionable data and half-baked ideas in poorly written journal articles.

CHOOSING A JOURNAL

The number of technical journals has jumped dramatically during the last decade. That means that no matter how narrow or special-ized your field of expertise might be, you have a choice of journals to which you can send your article. The choice is an important one since journals vary greatly in prestige. Most tenure decisions, for example, are based only on "reviewed" articles; that is, articles sent to the most prestigious journals that send out manuscripts for extensive review before they are accepted for publication. The journals also differ greatly in purpose and scope; if you submit a manuscript to the wrong journal, it will be rejected no matter how good it might be, and you will have thus delayed getting your information into print.

The only way to choose the best journal for your manuscript is to spend an afternoon at the nearest university or college library. From past reading in your area of expertise, you should have some idea of which journals publish articles relevant to your proposed manuscript. Carefully examine the contents of recent issues of several of these, checking, for example, the length, complexity, and orientation (theoretical vs. empirical) of typical articles. Also look at the names and affiliations of authors of recent articles to see if the journal attracts submissions from researchers and laboratories that you know well.

After you have chosen the most appropriate journal, copy the "Instructions to Authors" information, which is usually inside the front cover of every issue. That page will give you specific, detailed information on submission procedures for that particular journal. Follow it (and the supplementary information in this appendix) closely to maximize the chance of your article being accepted.

ASSEMBLING THE MANUSCRIPT PACKAGE

Once you have chosen a journal, follow these ten steps for producing and sending the manuscript.

1. Print the manuscript on bond paper using the best-quality typewriter or computer printer available to you. Double-space all text, footnotes, and references. Carefully follow the instructions for authors on handling reference citations since almost every journal's policy is different. Make enough photocopies of the manuscript so that you have one to keep after you send the required number of copies to the journal editor.

2. If your printer or typewriter lacks necessary scientific and mathematical symbols, use a black felt-tipped pen to create them. Make marginal notes if necessary to distinguish between Greek and standard characters. Underline everything to be set in italics.

3. Prepare a title page that includes the following:
   a. a title of ten to fifteen words
   b. a running title of three to five words
   c. the names of the authors and the affiliations of each

d. the address to which all correspondence concerning the manuscript should be sent

4. Prepare a list of three to five "key words"; that is, words by which the article should be indexed. Choose these carefully, since many readers will find your article only through information retrieval services such as the *Science Citation Index* that reference your article according to these key words.

5. Prepare an abstract of 100 to 250 words. Most journals print an article's abstract at the beginning of the article, but you should include one as a service to editors and reviewers, whether or not a particular journal requires an abstract.

    The abstract for a theoretical paper should describe the issue, the analysis, and implications for further research. The abstract of an experimental paper should describe the motivation for the research, the object, the procedures (briefly!), and the results and conclusions. Clearly, you will have to use subordination and a concise style (see Chapter 3) to pack all that information into 250 words.

6. If you have complex tables for inclusion, consult one of the style guides listed below for information on how to proceed. Because tables are so difficult for printers to typeset, and editors are sometimes reluctant to include them, do your part to make sure that the table is carefully and logically constructed (see Chapter 4), and typed in a way that will simplify the work of the printers. Include a table number and a brief, descriptive title at the top of each table. Refer to each table in the text.

7. Unlike tables, figures (that is, line drawings, photographs, charts, and graphs) are not redone by the printer. The author must supply quality, contrasty, glossy photographs of each figure that your manuscript calls for. How to produce these photographs and the original artwork is beyond the scope of this short book; our advice is to find a graphic artist who is experienced in preparing technical illustrations for publication.

    Work closely with that artist to achieve the goals for effective visuals, described in Chapter 4. Be sure to show the artist sample figures previously printed in the journal in which you are planning to publish; he or she will need to know the column width, quality of paper on which the journal is printed, and typical size and style of lettering. Have the artist make one eight-by-ten-inch print of each figure, and one set of smaller prints (the size that you wish each figure to appear in print) for each copy of the manuscript submitted.

8. Identify each figure by lightly writing the figure number and running title of the article on the back of each print. Check your manuscript to make sure that each figure is mentioned in the text. Type on separate pages a brief figure caption (double-spaced) for each figure. Usually the figure number and a short noun clause is sufficient:

> Figure 1. Locations of testing sites proposed by the Nuclear Regulatory Commission.

If a reader needs further information, such as an explanation of abbreviations or what to look for in a photograph, include that information in the caption. Finally, if the figure is adapted from information from another published source, the caption should acknowledge that source.

9. Write a cover letter to the journal editor to whom you are submitting the manuscript. Include the name of the journal to which you are submitting the article, and the full title of the article. You should also state that the article contains only original material, and that the manuscript is not being considered for publication by any other journal.

10. Assemble the original manuscript, extra copies as requested in the instructions for articles, eight-by-ten-inch prints of figures, figure legends, and cover letter. If you wish, include a self-addressed postcard that the editor can use to acknowledge receipt of the manuscript. Take care to protect the surface of the prints with tissue paper. Back the whole package with rigid cardboard to prevent bending of the prints in the mail.

REACTING TO REVIEWERS' COMMENTS

Since many journals have a backlog of articles, and the review process may be a protracted one, you may not know your article's fate for four to six months. It may be rejected, sent back for revisions, or accepted for publication as submitted.

If your manuscript is rejected, you should receive reviewers' comments stating why. You should do your best to revise the manuscript in view of those criticisms, have it reviewed by colleagues whose judgment you trust, and submit it to a different journal.

If, as is most common, your manuscript is accepted but returned for revisions, do your best to make revisions which eliminate your reviewers' objections. If a reviewer makes a comment that reflects lack of understanding, clarify as best you can the point that was misunderstood. However, if a reviewer suggests a change that is incompatible with your interpretation of the facts, then make no changes; it is, after all, your manuscript. Before you return the revised manuscript, write a cover letter to the editor explaining point-by-point the revisions you made in response to some of the reviewers' comments, and your reasons for not making changes in response to other comments. With luck, you should see your article in print in less than six months.

If you have the rare good luck to have your manuscript accepted for publication as submitted, write us a letter; we'd like to know your secret for success.

CORRECTING PROOFS

About a month before your article actually appears in print, you will receive from the journal typeset proofs for you to correct. Since the journal may not double-check for typographical errors, you should enlist the help of the most meticulous, painstaking fussbudget you know to help you in the proofreading.

Read aloud from the manuscript as the fussbudget reads from the proofs. Be extra careful to check titles, headings, figure captions, and tables, word for word and numeral by numeral. Use standard proofreader's marks (see, for example, *Webster's Ninth New Collegiate Dictionary*). Mark your corrections using a soft lead pencil. Make a photocopy of the corrected proofs before sending them back to the printer.

REFERENCES AND FURTHER READING

Barclay, William R., M. Therese Southgate, and Robert W. Mayo, Comps. 1981. *Manual for Authors & Editors: Editorial Style & Manuscript Preparation.* 7th ed. Los Altos, CA: Lange Medical Publications. (Available through Appleton & Lange.)

CBE Style Manual Committee. 1983. *CBE Style Manual: A Guide for Authors, Editors, and Publishers in the Biological Sciences.* 5th ed. rev. and expanded. Bethesda, MD: Council of Biology Editors, Inc.

The Chicago Manual of Style. 1982. 13th ed. Chicago: University of Chicago Press.

Form and Style Manual for ASTM Standards. 1986. 7th ed. Philadelphia: American Society for Testing and Materials.

A Manual for Authors of Mathematical Papers. 1973. 8th ed. Providence, RI: American Mathematical Society.

Meador, Roy. 1985. *Guidelines for Preparing Proposals.* Chelsea, MI: Lewis Publishers, Inc.

Publication Manual of the American Psychological Association. 1983. 3rd ed. Washington: American Psychological Association.

Webster's Ninth New Collegiate Dictionary. 1987. Springfield, MA: Merriam-Webster, Inc.

Appendix 4

Documentation

Documentation refers to the practice of providing your readers with references to your information sources. A reference can be as simple as a word or two in parentheses next to a graph or as complex as a full-blown bibliographic citation. Its purpose is twofold: to give your audience a place to go for further information about the topic and to avoid plagiarism. Both are important, but the first is merely a kindness. The second is an ethical and legal imperative, so we will take a closer look at it.

Plagiarism means claiming someone else's work as your own. Obviously, if you swipe a scholarly paper off your colleague's desk and type your own name at the top before submitting it to a journal, that's plagiarism. But it is equally plagiarism if you steal his rough draft and edit it before sending it in—or even take his outline and develop a paper from it. In each case you have used someone else's work and not acknowledged it. Although plagiarism may be unintentional, it is never blameless, because it is so simple to avoid.

The key to avoiding plagiarism is to document your sources. To do so means you have to be careful and systematic in your note taking as you gather information and also that you have to be careful and systematic in listing your sources in the final document. Note that you may write reports for which you have no outside sources: all your information is from your own testing or observations. Naturally, you need not list sources if there aren't any, but if you do draw information from others' work, be sure to provide references.

Several systems for listing references are in use, with a multitude of variations on each. At one time or another you have probably seen (or used) the footnote and bibliography system or the numbered list of references, to name just two. This book recommends a third system, the author-date system, because it is adaptable to both short and long documents, it is simple, and it is recommended by the *Chicago Manual of Style* (Chicago: University of Chicago Press, 1982), which is used as an authority by many publications. (Naturally, if you are submitting an article to a journal, you should consult that journal's style guide and use whatever form it dictates.)

In the author-date system, when you refer to a work in the text, you note it by listing the author, year of publication, and page number(s) in parentheses. For example, you might write something like this:

> Jones uses three arguments to bolster her assertion that superconductivity at normal temperatures will be attained within five years (Jones 1987, 357–359).

Then, at the end of the paper, you would list Jones's article along with the rest of your sources (in alphabetical order by author's last name), with a full reference:

> Jones, Alexandra T. 1987. Superconductivity in our time. *Journal of Science Predictions* 64(4):356–362.

The numbers following the title give the volume, the issue, and the pages on which the article may be found. Note that only the first word in the title is capitalized (if the title contains a colon, the word following the colon is also capitalized). If Ms. Jones had junior authors contributing to the article, their names would be listed after hers, but in normal order (first name first). Inverting the first author's name makes it easy to find in an alphabetical list; some publishers prefer inverting the names of all the authors of a work, but the *Chicago Manual of Style* recommends inverting only the first. Had Ms. Jones written a book instead of an article, it would be listed as follows:

> Jones, Alexandra T. 1987. *An Introduction to Superconductivity.* New York: Chalk Circle Publications.

For the myriad variations on these two basic situations (books with several authors, translators, anthologies with editors, second editions, and so on) you have two choices. You can either consult the

Chicago Manual of Style, which devotes more than 100 pages to the subject of documentation, or you can simply adapt these guidelines to your situation, making sure that whatever you come up with is consistent and clear and provides all the essential information. For a book, this means author, title, city and date of publication, publisher, and any special information, such as edition, translator, and so on. For a journal article, essential information is author, title, journal title, volume and date of the journal, and the pages on which the article appears.

The decision whether to abbreviate the name of the journal depends on your audience and your publisher. If you are sure that most of your audience would recognize the abbreviation, using it saves space. If you have any doubts, however, spell it out or include a list of abbreviations. As always, if you are submitting an article to a journal, follow its style guidelines.

Documentation of sources is a simple (albeit sometimes tedious) process, but it is an essential part of good scientific or engineering work. Clear, complete documentation adds to the credibility and usefulness of your reports.

Index

175